Praise fo

'Glaciologist Marco Tedesc
journalist Alberto Flores d
job of evoking Greenland's ice sheet.
National Geographic, Best Travel Books of 2020

'When it comes to ice and snow, and to understanding
the cold and exotic climates of Greenland and Antarctica,
Marco Tedesco is one of the most knowledgeable scientists
alive. *Ice* is a terrific complement to his important
academic work — a book that is insightful, lyrical,
and personal, and that will help guide readers through the
science of a warming world.' *Jon Gertner, author of* The Ice
at the End of the World *and* The Idea Factory

'As Marco Tedesco explains, the Greenland ice sheet plays
an oversized role in life on earth. Tedesco and Alberto
Flores d'Arcais have done a wonderful job evoking the ice
sheet's magnificence and fragility.' *Elizabeth Kolbert,
Pulitzer prize-winning author of* The Sixth Extinction

'Tedesco shares scientific and historical insight into
Arctic ice, discussing the hardy microorganisms that live
in it, the famed explorers behind major discoveries about
the ice, and the threat posed by climate change.'
Publishers Weekly

'Imagine a science book that's truly informative
but without an intimidating slew of equations, graphs,
and references. That book is *Ice* — a book about ice,
climate, Greenland, and the daily life of scientists who
study these esoteric topics. It's a beautifully told story
that will make you wish you could spend a year alone
on the Greenland ice sheet.' *Henry Pollack, Professor
Emeritus at The University of Michigan Department
of Earth and Environmental Sciences and author of*
A World Without Ice

ice

TALES FROM A DISAPPEARING WORLD

MARCO TEDESCO

with **Alberto Flores d'Arcais**

Translated by **Denise Muir**
Foreword by **Elizabeth Kolbert**

HEADLINE

Originally published in Italy as *Ghiaccio* by il Saggiatore in 2019.

First published in English in North America in revised form
by The Experiment, LLC, in 2020.

First published in Great Britain 2020 by
HEADLINE PUBLISHING GROUP
This edition published by arrangement with The Experiment, LLC

First published in paperback in Great Britain in 2022 by
HEADLINE PUBLISHING GROUP
This edition published by arrangement with The Experiment, LLC

1

Cataloguing in Publication Data is available from the British Library

ISBN 978 1 4722 7427 4

Offset in 10.18/16.65 pt Freight Text Pro by Jouve (UK), Milton Keynes

Printed and bound in Great Britain by Clays Ltd, Elcograf S.p.A.

Headline's policy is to use papers that are natural, renewable and recyclable
products and made from wood grown in well-managed forests and other
controlled sources. The logging and manufacturing processes are expected
to conform to the environmental regulations of the country of origin.

HEADLINE PUBLISHING GROUP
An Hachette UK Company
Carmelite House
50 Victoria Embankment
London
EC4Y 0DZ

www.headline.co.uk
www.hachette.co.uk

For my daughters, Olivia and Francesca,
rivers of light

To Alice

Contents

Foreword

by ELIZABETH KOLBERT

I visited Greenland for the first time in the spring of 2001. I'd recently read a book about the Greenland ice sheet and the millennia of climate information frozen into its two-miles-deep expanse. Scientists were attempting to drill to the bottom of the ice sheet to get at snow that had fallen 130,000 years ago, before the start of the last ice age. The project struck me as—forgive the pun—unbelievably cool. As a journalist, of course, I decided to write about it. It turned out that the New York Air National Guard, which services scientific expeditions to Greenland, also occasionally takes reporters. I wangled a seat on the next available flight. Twenty years later, I can still vividly remember landing in Kangerlussuaq, on the island's west coast. Several musk oxen were grazing by the airport. The following day, I took a ski-equipped C-130 to the drilling operation at the top of the ice sheet. I was hooked.

"Greenland!" the artist Rockwell Kent wrote, after being shipwrecked in an ice fjord. "Oh God, how beautiful the world can be."

Since then, I have made two more trips to Greenland, and each time I've been fortunate enough to get back onto the ice sheet. The experience has always been amazing—as close as a person can get to visiting another planet. But the reason I keep going back, aside from the fantastic beauty of the place, is the extraordinary significance Greenland has for life on this planet, which is the only one we are likely to be granted.

As Marco Tedesco explains in the chapters that follow, the Greenland ice sheet plays an oversized role in life on earth, particularly when it comes to global sea levels. The ice sheet is a holdover from the last ice age, when mile-high glaciers extended over vast stretches of the northern hemisphere. In most places—Canada, Scandinavia, New England, the upper Midwest—the ice melted away about 10,000 years ago. In Greenland, it has, so far, persisted. But there are many signs—signs that Tedesco and his colleagues are monitoring—that this is changing.

Greenland is losing ice, and the rate at which this is happening is accelerating. Were the entire ice sheet to melt away, global sea levels would rise by twenty feet, and most major coastal cities would be inundated. The process could take centuries to fully play out, but in the meantime, sea levels would rise by several inches a decade. The rise wouldn't be evenly distributed because, as Tedesco points out, the Greenland ice sheet is so massive it exerts its own gravitational pull, tugging water toward it. As the ice sheet melts, this pull will weaken, and so, counter-intuitively, the rise will be greater farther away from Greenland's shores.

I'm not sure I will have the chance to visit the Greenland ice sheet again. This makes me sad to think about, as it is perhaps my favorite place on earth. But flying off to see the ice sheet only helps hasten its demise. Tedesco and Alberto Flores d'Arcais have done a wonderful job evoking the ice sheet's magnificence and fragility. I think in the future I may be better off staying home with a good book, like this one.

ELIZABETH KOLBERT has been a staff writer at *The New Yorker* since 1999. Her journalism has garnered multiple awards, including a 2006 National Academies Communication Award for her three-part series "The Climate of Man," which investigated the consequences of disappearing ice on the planet. She is author of *The Prophet of Love, Field Notes from a Catastrophe,* and *The Sixth Extinction,* for which she won the Pulitzer Prize for general nonfiction in 2015. She received the Blake-Dodd Prize, from the American Academy of Arts and Letters, in 2017.

Prologue

It's dawn. The sun rises in front of the Montevergine, bathing the mountain in light and dispersing the mist hanging in the valley that cradles the sleeping city of my childhood. This part of Irpinia lies still between the peaks of Partenio and Terminio, two Colossi of Rhodes of this area that stand guard over a culture and a place that are unchanged by time. These are not the immense mountains of America, which are impossible to fully behold; neither are they the Dolomites, with their unattainable, impenetrable heights jutting mysteriously into the clouds. These are mountains on a human scale, a dimension that has kept intact the flavor of the land, the smell of its pastures, and the harshness of its landscapes. A harshness that callouses hands and toughens the spirit, and that has forged the thoughts and actions of a proto-European people— the Hirpini (from hirpus, Oscan for wolf); 90 percent of my gene pool originates from this place.

These were the mountains I dreamt of climbing as a child, impatient to get to their summits, to challenge both myself and the mysteries they held. I have no doubt this is where I learned the slow "geological" pace that affords me time to absorb my surroundings, to observe, to contemplate the sights before me and assimilate them within me. Neither the sea in Naples (where I moved to study engineering) or Brazil (where I have lived on several occasions) ever subsumed or weakened this bond. When I used to gaze, in awe, at my local mountains, eyes sweeping up from base to summit, I never imagined I would one day set foot on the glaciers of Antarctica, walk in the Rocky Mountains, trek the Alaska plains, explore the forests of Finland, or tread on volcanic rock lying over ice in Iceland.

Despite being from southern climes, when I started my PhD, I found myself studying something that belongs to a very distant world: first snow, then ice. It was in those years—now almost two decades ago—that I made my first visit to the glaciers of the Dolomites and felt irresistibly drawn to the endless, solitary expanses of white as I looked out at them from high up on the summit. In awe at the dazzling landscape before me, the irrevocable, unalterable decision to see the ice sheets of Greenland for myself was already firmly cemented in my head and heart. Several years later, this decision was to become a reality, and little by little, year after year, one expedition at a time, that same country became part of my life. It was the start of a long and personal voyage of exploration of a world that, after so many years, never ceases to astonish and to fascinate me—as it did the very first time I laid eyes on it.

It must've been fate.

CHAPTER 1

The Origins of Ice

'M FIRST TO WAKE UP. I often am. The silence surrounding me is absolute.

There's something special about night in the Arctic. I'll never forget the first time I slept here: the emotion of being camped on the immense ice sheet, the light of the sun that never sets, a constant companion for people in this job. I've always woken up early and find it almost impossible to go back to sleep once I'm awake—a habit that intensified when I became a father and has never left me since.

The first "trial" of the day when living on the ice is to get dressed. It's not as easy as you might think. To prepare for the world outside the tent, you need to be wearing multiple layers. Some call it dressing like an onion: You don different layers of differing weights, each one serving a different purpose—for example, a base layer closest to the skin, intermediate layers, then an outer layer to provide protection from the wind.

It takes a serious degree of body contortions: The tent is only a meter and a half high, so every movement has to be carefully coordinated. I rock on my back to pull up my trousers, sit cross-legged for upper body layers, and then attempt to "slide" on two pairs of thick socks, which is difficult because my now-exposed feet are usually like blocks of ice by then. Then there's the golden rule: No cotton. Clothes keep us warm because of the way they trap warm air next to the skin, but the problem with cotton is that once the clothes get wet, the pockets of trapped (warm) air in the fabric fill up with water and don't protect you anymore. Like how, when you're out walking and start to sweat, a cotton top will soak up the sweat like a sponge and make you feel cold, since your initial body temperature was warmer than the temperature of the air outside (which is very much the case in Greenland). Once cotton clothes are wet, they no longer provide the necessary insulation, which is why we always make sure we bring things made of insulating fabric—either wool or a synthetic material.

I scooch over to the zipper on the tent door. I have to be extra careful not to wake my travel and research companions: Anywhere else, the metallic whoosh of a zip opening would be almost inaudible. Not here. On the ice, even the tiniest of noises is amplified. Our tents are of the standard camping variety, known in the trade as "four-season." They're light and go up in less than twenty minutes. The outer layer is waterproof to protect us from the rain, which does actually fall in Greenland. Many people think it's cold inside the tents but, generally speaking, that's not the case. When the sky's clear, the hot Greenland sun warms them up so much we have to open them at night to let some cool air in before we go to sleep. It's worse in the middle of summer when the sun never sets.

As I mentioned before, there's an unearthly hush around our camp; the only source of noise pollution, if we can call it that, is the whistling of the wind, which sometimes blows in a steady stream, sometimes in bursts. I finally manage to pull the zipper down and, to my ears, it sounds nothing short of an explosion. That's to be expected, though, when you think that sound is the transmission of pressure waves arriving in the ear and encoded by the brain. Since the air is so rarefied in Greenland and there are no other sources of noise, ordinary sounds are perceived in a completely different way and with an entirely different timbre. Or maybe it's exhaustion and auditory hallucination, or just the cold playing tricks with our senses.

I crawl out on my knees, pull myself onto the waterproof mat outside the door, and sit up. Time for one last effort: to get my boots over the woolen socks that are too thick, but essential. I'm tired already. Tired, yet excited at the prospect of what lies ahead, aware that we only have the objects we brought with us to overcome every obstacle and to deal with every situation that might present itself. When you're in the middle of the ice sheet in Greenland, there's no stopping by the local supermarket or hardware store if you're missing a screwdriver or need a ball of string.

I can't tell if the others are awake: The only sound coming from the tent is the rhythmic susurration of people breathing. Last night was one of the kind I call "interesting"—when someone wakes up and then wakes *you* up to fire a question at you, talk about an idea, or—as happens more frequently—is alarmed because of something they think they heard. It was Patrick's turn this time.

Patrick was one of my PhD students—he had never been out of New York and had gleaned his knowledge of Greenland purely from satellite data and models. I invited him to join our team to give him a well-deserved opportunity to grow further as a professional in the field, and also to give him firsthand experience of the ice. I truly believe that if you're studying this immense and magnificent ice mass, you have to see it for yourself, at least once in your life. It must've been around three in the morning when Patrick woke me. He seemed on edge and wanted to know if I'd heard the loud noise, like a thunderclap, from the ice below us. "Whatever you need, just give me a shout, even if it's the middle of the night," I'd glibly reassured him when we'd landed. He'd taken me at my word.

I tried to put his mind to rest, explaining that the ice often creaks and that given the absence of any other noise around us, it's easy to imagine things. What we normally hear, though, are muffled thuds, like something splitting beneath us, the noise an enormous stone would make when it lands with a thud on mountainous terrain. I suggested he go back to sleep, with the reassurance that there was nothing to

worry about. Not that I was wholly convinced, either; clearly, when you're deep in the Arctic you shouldn't rule anything out, no matter how small and seemingly insignificant.

A few minutes after my chat with Patrick, I heard it, too,
the noise he'd been talking about. It was the ice, flowing
powerful and inexorable beneath us at speeds, in summer,
of up to about half a mile a day on the top layer. In lay-
man's terms, that would be like pitching your tent outside
the Empire State Building at night and waking up the next
morning in Madison Square Park. Patrick had touched a
nerve, though, and I couldn't get back to sleep, partly because I was worried and partly out of excitement. My mus-
cles were tensed, ears pricked, waiting for a sound, any
sound, however small and inaudible—like I was listening to
a dinosaur's breathing with a stethoscope, only in my head.

Not many people know that the ice moves; there's
a common perception that the ice in Greenland is static, motionless—inanimate, in other words—when it's
anything but. *Panta rhei*, the ancient Greeks taught us:
Everything flows. Even ice masses, like dense rivers, move
under their own weight. The flow slows in winter, when
it's colder and the ice less fluid, but in summer it's like
driving downhill on a wet road: No amount of braking can
stop the forward motion. In the "warm" season, meltwater
seeps through cracks and fissures in the ice, accelerating
flow rates when it reaches the rock below the sliding ice.

I think about this as I look down at my boots, feet
tucked safely inside. These aren't the boots I'll wear for

our excursion later—they're knee-high and ideal for camp life but not for trekking. The padding inside protects your feet at temperatures of up to –104°F (–40°C). Not my feet, though. They're having none of it. They're cold from the minute I leave the tent in the morning until I get back at the end of the day. "You must be used to the cold by now," people often say to me, "you probably don't even notice it." Actually, the opposite is true: I have a fairly slender build and don't have the mass I need to retain body heat. On excursions, we wear either hiking or Arctic crossing boots, which are sturdier and provide more ankle support to protect us from sprains. But the flip side of this is that they offer slightly less protection from the cold.

Once outside, I sink into the folding chair near the tent door and think how perfect a hot cup of coffee would be right now . . . but also how I should probably wait until everyone else wakes up. My mind drifts again, as if I'm still half asleep. I think about how impossible it is to quantify, or put a price on, the good fortune I have—I can think of no better way to describe it—to be looking out at this stunning landscape: in such profound stillness, surrounded by snow and ice.

If you've never been to the Arctic then there are a few things that would no doubt surprise you on your first visit. Even just at first glance. What I'm looking at now is anything but mundane or flat: Snow dunes, a few meters long, desert-like, line up in parallel with the wind direction. You have to pay attention to things like this when putting up the tents. If you get them in the right position with respect to the wind, it can stop them from filling up with snowflakes, glistening and lining the tent floor with a fine sheet of delicate diamonds. The glimmer reminds me of the huge waves I saw surfers ride in Hawaii or crashing on the beaches of Rio de Janeiro. The sparkle comes from the snowflakes fracturing when they hit the ground, split by the wind and other factors into tiny shards that then scatter randomly. The snowflakes, or what's left of them, behave like a multitude of minuscule mirrors strewn over the surface, reflecting the sunlight in all directions. That's why they seem to glimmer.

In awe, I continue to gaze out. A veil of wind-borne snow is painting circular patterns across the surface, reproducing the gusts and whirls of the glacier breeze. It's as if there's an artist standing behind me, adding details to a painting as I watch—someone who has dotted color at various points on the horizon, then decided to fan it out using their brush like a spatula. The turquoise sky— uniquely turquoise on account of the thinner air and lower humidity than southern latitudes—provides a backdrop for my meandering mind. It's a majestic color, and despite lacking the force of rain-packed clouds over the English countryside or the sudden violence of an equatorial storm, it has a quiet power—like a gigantic wave of color. It is majestic yet motionless, aware of its own vastness and strength, but with no need to flaunt it.

The enormity of the sky is determined by the vastness of the mass it embraces. The two elements—sky and ice—share the same space, totalitarian in their chromatic supremacy, leaving no room for anything but blue and white. I have a flashback. Images spring to mind of when I was flying back in a helicopter from my first trip to Greenland, after endless weeks of white, when I saw the green, the reds, and the full rugged tundra color palette again. That's when I realized—when I felt—as if my time on the ice had been spent speaking and hearing only a few words. As if someone had cut away part of my chromatic vocabulary, and part of me with it.

The desolation of the ice before me is a pleasant one, infusing me with a sense of peace and tranquility. Still halfway between sleep and wakefulness, I let this feeling wash over me, like I'm bobbing in a lifeboat on a calm sea. Time here is different: It's more like the "geological" time of my homeland. I don't need a watch. I can't and don't need to catch up on the latest news or download new music. The passage of time, as we normally think of or experience it, has no meaning here.

The ice is an elephant; I am a mere cell. Here in Greenland, it has taken thousands of years to form: Snow falls, year after year, never melting in the summer, becoming buried under yet more layers of snow. Greenland was created and has developed through the action of millions of minuscule particles—snowflakes that pile up with the passing years, snow layers compressed under their own weight, forcing out the air until they turn into ice with that all-important density, a dogma for scientists like me: 917 kg/m^3. Nine hundred and seventeen. The magic number. Less than 10 percent of the structure of glacial ice is air; all the rest is water in a solid state. Ice formation is a constant process that has been taking place for decades, centuries, millennia. When ice reaches a "critical" mass, it begins to flow, literally, under its own weight. Gravity, that mysterious natural force, sculpting the world around us yet again.

It takes much less to melt the ice: Nature's slow, painstaking labors can be reversed in a day, maybe even less. The vast, icy expanse moves at a natural pace, heedless of us, of contemporary society, of humans, dashing around like crazed cells trying to take everything in before the next digital distraction—the next screen—grabs our attention. We human beings are tiny, but with our greenhouse gas emissions and global warming, we're like a virus attacking everything and everyone—and have managed not only to threaten the mighty Greenland but also to bring it to its knees.

At the center of this polar landmass, at its thickest point, the ice can be up to 2 miles (3 km) deep, thinning out to around a few hundred meters along the coast where it flows into the ocean, like a river of opalescent lava. The deepest sections of ice are also the oldest and the ones that have been compressed under the most weight, resting on the granite bedrock for thousands of years. Little by little, as the ice slides toward the ocean, preparing to return to its origins, it melts on the surface, releasing a part of its memory into the ocean. Layers of ice deposited at different times become distorted, undulated, fused together by the constant movement, and the surface ultimately merges with the deeper roots.

I go over the last few details of our experiments in my head, reminding myself of the things I must do and those I must not. We're preparing to collect data that will help us understand the impact of climate change on the rate of the ice's melting and, in turn, what role this plays in rising sea levels. We'll be studying not only the impact of rising temperatures on the formation and evolution of the systems of meltwater rivers and lakes, but also how the sun—and the fact that ice is getting "darker"—are a key part of this. We know, though, that Greenland is more than all this—and another reason we're here is simply to savor it, to absorb it, and to get to know it.

I gaze and dream, dream and think. I think back to the southern Italy of my youth: a rich, harsh land where roots run very deep for those who still live there or for those who maintain strong bonds despite living elsewhere for many years. My roots are the linchpin of my existence: They define me and influence me, even in the US, my adopted home, the place I've hung my hat for several years now. Roots provide an anchor—restraining, circumscribing, restricting. They give us stability, but they also sometimes make it more difficult to effect real change. They shape us slowly, bit by bit, as if mimicking the gentle rhythm of sap bringing forth a tree from the inert soil—nurturing its branches, growing its leaves.

The sluggishness of the morning, the accumulated fatigue, and the longing—now more a need—for coffee, make me feel hallucinatory for a moment. I imagine roots reaching deep into the icy ground, roots that move, following the flow of the ice and the flow of time. On this cold morning in the Arctic, I feel at ease with this train of thought: After all, I also moved a long way from my place of birth, a move that was necessary for new branches and new leaves to sprout—the lifeblood of their roots giving them the strength they needed to grow and mature, the spores and seeds around them helping them grow stronger and reach further.

In this way, ice, and its static vitality, reminds me a little of myself. It reminds me of all of us.

CHAPTER 2

Mother Greenland

THE WATER FOR THE COFFEE is nearly boiling. I open the airtight tin holding the precious ground beans, and the pungent aroma fills the air inside our kitchen tent instantly, helped by the lack of competition from any other odors, unlike in the city. I spoon coffee into the French press and fill it with water, and brown bubbles foam on the surface; I put the lid on, wait a few minutes, then push on the plunger, forcing the mesh disc down to the bottom, separating the grounds from the fragrant black liquid.

The others join me: Patrick; Ian and Alison, who are British, close friends and hydrology experts; then Christine, who's studying extreme life on glaciers. Coffee is a kind of ritual, or "the poetry of life," as the great Eduardo De Filippo described it in the legendary balcony scene in *Oh, These Ghosts!* We have a selection of things we brought with us in the helicopter to eat with our coffee: powdered milk to mix with more water, cereal, dried fruit, walnuts, and an assortment of options that include brown sugar and blackberry jam.

The further we get into our expedition, the more we'll have to get used to a less sophisticated, more frugal breakfast, concocting dishes that are scientific experiments in themselves. As the first trickle of coffee scalds my tongue, I look outside and see the sky change from cobalt to cerulean blue, the sun rising over the horizon (not that it ever fully disappears below it) and starting its rotation over our heads as if anchored to the earth by an invisible thread that stops it from getting away.

The sun—everything depends on it: our climate, life on planet Earth, the melting of Greenland and the rising sea levels. The sun—and through it the temperature of our planet, which has been rising disproportionately and gradually, transforming our beloved Earth into a place that may soon become uninhabitable for human beings.

Brother Sun and Sister Moon: so closely associated in our culture with the concept of man and woman. But not in Greenland. Here in the land of ice, the sun is female and the moon is male. While the others finish breakfast, I pull out my laptop and pick a few articles to read from the many I downloaded before I left, in particular the ones on Greenlandic culture. Christine and Ian ask what has me so absorbed, so I read it aloud to them.

In Inuit legend, Malina (the sun) and Anningan (the moon), sister and brother, respectively, were very close as children but later grew apart when they were sent to

live in separate lodges for women and i.
Anningan watched the women, he realized .
the most beautiful of them all, and he took n
unique material of her clothes. One night he went
her in her bed and, taking advantage of the dark, m.. ı
with Malina, who had no idea the man was her brother.
But when he returned a second time, Malina was ready:
She smeared soot from the oil lamps on her hands so that
it would rub off on the face of her abuser so she could
identify him the next day. She waited patiently for Anningan to return, and when he did, she touched his face and
left it black with soot.

The next day Malina began her search for the rogue and,
to her great sorrow and surprise, realized the guilty party was Anningan. In a blind rage, she cut off her breasts
and offered them to her brother, roaring, "Since you seem
to like me so much, eat these!"; then she fled. Anningan,
equally enraged, chased after her—and they both ran so
fast they rose off the ground and became two bodies that
continue to chase each other in the sky.

The legend says that Anningan often forgets to eat
while chasing his sister, getting thinner and thinner until
he disappears for three days every month (the period it
takes for a new moon to emerge) in order to revive himself, recoup his strength, and resume the chase. I stop
reading and, when I look up, see the others gazing at me,
bewildered. Yes, it's a powerful legend that evokes violent
images and violent events, but it can also be considered

an emblematic example of the deep and visceral bond the local people have with the natural world.

Silence returns, and it feels a bit like I'm sitting around the kitchen table at home. We're halfway through breakfast and there's still some time left before we need to pack our bags for the expedition. I've always been fascinated by stories and legends, especially folklore, so I decide to read up on Inuit mythology. A number of indigenous myths recount how the people here scrabbled to understand the world around them; two in particular capture my imagination: the story of Akhlut and the story of Nanook.

Akhlut is an orca spirit that takes the form of a giant wolf or wolf-orca hybrid. The myth describes Akhlut as a vicious beast that ventures onto land to hunt humans and leaves wolf tracks to and from the sea, signaling that it lies in wait for unsuspecting prey under the water. This story explains why the townsfolk here believe that if huskies (sled dogs) go near or into the water, they have been possessed by the devil.

Nanook is the master of polar bears. Inuits believe him to be powerful and mighty—and almost human, since polar bears' tracks make them look like they walk upright, though in reality they're placing their back paws in the prints left by their front paws to get across the snow more easily and without sinking. Hunters tend to worship Nanook because they believe he determines their success. Indeed, the Inuits think polar bears will let themselves be killed to obtain the gifts hunters offer their spirits after death. If a male bear is killed, the hunters leave weapons and other hunting tools on the ice, whereas needle cases, scrapers, and knives are proffered for female bears. The legend has it that if a dead polar bear is treated honorably by the hunter, it will share the good news with other bears who will then be eager (so to speak) to be killed by that particular hunter.

Ian wakes me up from my reverie about the moon chasing the sun and a wolf-orca trying to savage them midflight. He gives me news of a polar bear sighting. It's not the "usual" sighting in a local village or community (now reported more frequently due to climate change) but something much more out of the ordinary: In living memory, a polar bear has never been seen so far from the coast. The encounter, if we can call it that, took place at the Summit observing station, located at the highest point of the island, at an altitude of about 10,000 feet (3,000 m) and a few miles from the coast. Summit is a permanent Arctic base camp, a scientific research site that regularly hosts scores of people, including both researchers and their logistics teams, with tents offering lodging for visitors and a main facility housing the kitchen and work spaces.

I'm suddenly reminded of the disorientation and light-headedness I felt during my last visit to Summit several years ago, because of the altitude and thinner air. Back then, we spent a week moving around on snowmobiles to collect data, and even though it was the height of summer, the temperatures were as low as −77°F (−25°C). According to this news, the Summit bear was seen "sniffing around" the visitor tents, sowing panic and distress among the camp's inhabitants. We look at each other in dismay as we finish off the last few morsels of breakfast: What can have pushed the bear so far in from the coast? What state was it in when it reached Summit? In a moment of empathy, we talk about how exhausting it must have been for the bear to make it all the way to the camp with nothing to sustain it, no matter how strong and mighty polar bears generally are.

My mind goes back to the days prior to our departure when some colleagues had been talking about the extent of sea ice the previous spring, that it was well below the seasonal average and had actually been a record low. When I glance at Ian, an English gentleman in his early fifties, I see from his expression that we're thinking the same thing. Maybe, he suggests, the bear's unexpected foray inland was a desperate attempt to follow the odors coming down from the station, carried on the wind that descends Greenland's coastlines. After all, polar bears are renowned for their amazing sense of smell, so powerful they can smell a seal under 3 feet (1 m) of ice or from more than a mile (1.5 km) away.

Christine, who has more biological expertise than the rest of us, shakes her head. She has a proud look in her green eyes and a no-nonsense manner, traits she has developed in order to rise through academia, a world in which women are still a minority. "That's impossible," she tells us.

Even with such a strong sense of smell, no animal would ever be able to pick up a scent from such a distance. Patrick looks up from his bowl of cereal, ginger beard dangling dangerously over the milk swaying from the sudden movement. His gaze sweeps upward from the crate he's perched on, one of those containing our equipment. He offers another theory: that the bear followed the smells left behind by the land convoys sometimes used to transport heavy materials and collect ground data. Or maybe, I think, the bear wandered onto the ice by accident, its internal compass went amiss, and it set off in the opposite direction from the coast, picked up the smells, and followed them up to the Summit camp.

We can't agree on an explanation and, being an enthusiastic group of scientists who seek answers for everything, we feel a little frustrated. Frustration turns to sadness when Ian tells us that after numerous attempts to keep the bear away from the camp, it was killed. I voice my disapproval and wonder why they didn't try to put the animal to sleep and transport it back to the coast. After all, the planes flying into the base are big enough to carry both the animal and a cage big enough for it. The only answer we can come up with is that it would've been too expensive and the life of bear was not deemed worth it.

The mysterious sighting of the polar bear at Summit triggers the memory of something else that happened at the other end of the world, in Antarctica. It was actually a mysterious *dis*appearance this time, of more than 100,000 Adélie penguins from a rocky point at the head of Commonwealth Bay called Cape Denison. The promontory is named after Sir Hugh Robert Denison, who funded the Australasian Antarctic Expedition led by Douglas Mawson a century ago, and who called the site "the windiest place on Earth."

In 2011, there were more than 100,000—possibly as many as 150,000—penguins in the colony, but now only a few thousand remain. If we're looking for a guilty party, then B09B, a colossal, 38-square-mile (100 km²) iceberg (not one of the robots in *Star Wars*), is to blame. After quietly floating around in the open sea for twenty years, in 2010 the iceberg became grounded off the coast of Commonwealth Bay and landlocked the penguins. No longer able to get to the ocean where they normally fished, the penguins were forced to trek up to 37 miles (60 km) to find food—and then make the same trek back to bring it to their young.

No one knows what actually happened to the penguins and why they disappeared. The worst-case theory is that a large part of the colony was wiped out by the difficulty to find food. The alternative, less pessimistic, and possibly more likely theory is that the penguin colony moved after the iceberg became jammed in the bay.

I shrug in silence. Who knows if we'll ever find out the truth?

We're full of energy now and no longer feel the cold. Well and truly wide awake. The conversation returns to the Inuits as we clear the table and wash our dishes. One of the first Inuit peoples to settle here were the Thule, who arrived on the Arctic coastlines around eight hundred years ago. They adapted quickly to life in such a harsh environment and advanced eastward across the Bering Strait to Greenland. Legend tells us they had heard about iron and the amazing properties that made it ideal for the construction of tools. They were convinced this magical material was to be found in the meteorites of northernmost Greenland, known classically as "Ultima Thule." Like the ancient Egyptians, the Thule people had discovered that the enormous stones falling from the sky contained a precious material that had the power to alter the history of humankind. The Inuits were possessive of their meteorites, and for good reason. When they showed them to Admiral Robert Peary, without asking their permission he loaded them onto a ship bound for New York, where they are now on display at the American Museum of Natural History. If you ever get the chance to go and see them, stop for a second just to consider that you're looking at the loot from a crime that was never punished.

The Thule people were mainly hunters of whales and seals and were probably the first to bring dogs to Greenland, marking the beginning of the tradition of sled racing that continues today. Dogs were used (then and now) to travel to hunting locations. Unfortunately, temperatures in the Arctic have risen over the centuries and continue to rise—twice as fast as the rest of the planet—endangering not only the environment but also the entire ecosystem and these traditions. The people of Greenland—the Saqqaq culture were its first known settlers, beginning four thousand years ago—also tell us that hunters have to get the seals they catch out of their nets as quickly as possible, otherwise worms and parasites they had never seen before will attack the carcasses and destroy them in a matter of hours. The hunters believe these vermin have traveled north from southern areas because of rising ocean temperatures.

Another threat the people of Greenland face is the impact climate change is having on their economy. Even smaller communities in northwest Greenland now have electrical energy, albeit provided in small part by diesel generators. In order to afford electricity, at least one member of every family must have fixed paid employment. In the majority of cases, that person is a woman, be it wife, daughter, or mother. The work they do means men can continue to go out and hunt all day. One of the consequences of this division of labor is that Inuit women are losing their proficiency at age-old traditions (like skinning and treating skins) much faster than men, who, by force of circumstance, are also spending more time with "modern" contraptions like snowmobiles and rifles instead of with dogs and harpoons.

As a lover of all kinds of music, I can't resist the temptation to slip Greenlandic musical culture into the conversation. Luckily, we're all on the same page for this one: Patrick is an extremely talented classical pianist; I play a number of instruments; and Ian, who we've discovered has the voice of a nightingale, will often sing us the occasional Queen song.

The Greenlandic Inuits share a strong musical tradition with their neighbors across the Canadian territories of Yukon, Nunavut, and the Northwest Territories—as well as the US state of Alaska. Greenlandic Inuit music is largely based around singing and percussion, and generally reserved for large celebrations and other important gatherings. The earliest recordings of this music were made in 1905, and it is still performed today. Greenlandic drums are made of animal skin stretched over a wooden frame and decorated with symbolic motifs by the drummer. The harp and fiddle are also to be found, most likely the result of interaction with other cultures. Greenlandic drum dances are based around a single dancer who improvises as he dances, usually in a *qaggi*, which is a snowhouse built for community events such as the drum dance. The musician's skills are measured by his endurance in his lengthy performance and the nature of his compositions.

Drum dances are an important element of Greenlandic Inuit cultural cohesion and serve as both a means of personal expression and pure entertainment: Generally lighthearted, convivial events, they can sometimes be competitive, with two contenders performing a singing and dancing parody, taunting each other and pointing out their opponent's flaws. This competition can often be used to settle feuds between warring families or individuals; the jokes are prepared ahead of time, and the person who evokes the most laughter from the audience is considered the victor.

As much as we appreciate the richness of Greenlandic Inuit tradition and all its joys, we also know there is profound sadness in these communities, too. Patrick has studied the social situation in Greenland and speaks to us of the scarily high suicide rate in this area (an average of eighty suicides for every hundred thousand people, which is more than twice the rate of the second-placed country, Lithuania). In reality, Greenland never appears in statistics because it's not considered a real country, just a region of Denmark, which has a much lower rate than in Greenland: only nine suicides for every hundred thousand people.

The number of suicides in Greenland began to creep up in the 1970s or thereabouts, mainly in cities where it peaked at 107 suicides for every hundred thousand people in 1994. It's not clear to what extent this tragic social drama is linked to geographical isolation, low population density, and the cold climate. On top of these factors is another worrying fact: The majority of Greenlanders taking their lives are teenagers and young adults. (In most other nations, elderly people dominate suicide statistics.) A 2008 survey found that one in four young women in

Greenland had attempted suicide. Some people claim that Greenland's young people are choosing to end their lives for the same reason they make a lot of their life choices: because their friends are doing it. This may be an oversimplified interpretation that trivializes the issue. We think about it in silence. I arrange my backpack and my tools and sit back down to wait for the others to be ready.

CHAPTER 3

The Color of Greenland

THERE'S A SUDDEN GAP IN the clouds that makes way for a blinding light to shine through. It's Greenland's ubiquitous summer sun, bright and limpid, a beacon of sorts, revealing neither east nor west, perennially orbiting over our heads in a gravitational slingshot. Let's not forget that the sun never sets in summer and never rises in winter on account of the tilt of our planet's axis, which exposes the Northern Hemisphere to the sun in summer and the Southern Hemisphere in winter. There's a simple experiment you can try to see this for yourself: Stick a knife into an apple, then shine a light onto the apple from the right side while holding the knife in your left, slightly tilted to the right; you'll see that only the top part of the apple is illuminated and the bottom isn't. Do the same again with the "sun" in your other hand, and you'll see the bottom part will be illuminated while the top will be in the dark. This is what happens with our planet. There's no way to get your bearings from the sun here, and our compasses are also pretty much useless. This is because magnetic north and geographic north are not the same, and the confusion between the two increases the closer you get to geographical north. Imagine looking out your window in Europe and pointing to the United States, using Canada as a reference point: The closer you get to the North American continent, the more inaccurate this would become.

We've just left base camp and are heading to the first of our data collection points on the "color" of Greenland. In the time it takes me to get my prescription glasses off and sunglasses on, I'm already blinded. You don't even need to look straight at the sun for this to happen: Just open your eyes quickly and it's as if a hundred camera bulbs are flashing in your eyes. The beam of light is so powerful and fast that my pupils can't contract in time to protect me from the dazzle. The areas around our camp, for the most part under a blanket of snow, reflect visible light by 360 degrees, and it feels like someone has just pointed a powerful laser beam in my eyes.

But despite all this, I feel lucky. If it had been "fresh" snow, formed of freshly fallen flakes, the energy hitting my eyes would have been far greater. It's a matter of simple physics: Fresh snow reflects around 90 percent of sunlight. To sum up, I'd say looking straight at the sun or staring at fresh snow on a sunny day are more or less the same thing.

Fresh snowflakes are small and wedge-shaped. Although it might not seem like it at first, they are very different from the larger, rounder flakes of older snow, which has already been through several melt-refreeze cycles. The amount of radiation reflected by the snowpack depends on how far tiny energy-carrying particles called photons have to travel: The shorter the distance they travel through the ice, the less energy is absorbed by the snow, and the greater amount of light reflected.

Luckily the snow around our camp—the snow we see when we leave our tents—is not that bright, although not everyone agrees. Christine and Alison point out that they can't see any difference between this snow and the fresh snow that fell a few days earlier. It's difficult in Greenland to draw such a clear distinction between perception and reality. I crack a smile, reminded of a popular bumper sticker I've seen Americans put on the back of their large, often monstrously large, cars. It's a popular saying of unknown origin: "Don't believe everything your brain tells you."

I have to say, I quite like it; it's clearly referring to the fact that the world around us has a huge influence on the way we think and that we must constantly challenge ourselves. For people like myself and the others on this trip, who are used to the rules (including the unwritten ones) of the academic and scientific worlds, it also has a more specific meaning: Never accept notions as dogmatic paradigms. Try to bring a sense of critical analysis to everything. In the scientific world, we test hypotheses and models; we do not seek absolute, irrefutable truths.

Either way, out here, even the most solid of certainties can falter. What our eyes see—the world we perceive as the reality around us—may not reflect what is actually the case, only what we are able to comprehend. There's one example that explains this better than any other: the color of the ice.

Let me share a personal memory with you. After my first expedition to Greenland, I went to see my sister, Samantha, only a few years younger than me, who appeared holding one of those atlases we used to have at school. She'd found it digging through the remnants of our adolescence, packed away in an old wardrobe along with our school textbooks, notebooks, drawings, and many more things that had been yellowed by time but remained vivid in my mind. It was an old atlas, printed when there was still such as thing as the Soviet Union, and a united Europe wasn't even a shadow on the horizon. Pointing to the whiter-than-white depiction of the island on the map, Samantha fired a question at me before I'd even had time to take my jacket off, "Is Greenland really that white? Shouldn't it be darker?" she said.

Samantha's question was anything but outlandish. Almost everyone thinks snow and ice are white, but the truth is, they're not. Not exactly. Greenland is covered in snow in winter, this much is true, and ice is definitely "lighter" than the forest or rocks, but it's still not white.

The color of Greenland is in its name—Kalaallit Nunaat, meaning "land of the people" or "land of the Kalaallit," the indigenous Inuit people who inhabit the west of the island, evolved over time into the Danish Grønland, translated as "Greenland." According to legend and historical sources, the name was the invention of Erik the Red, the famous Viking explorer—full name Erik Thorvaldsson—who lived at the end of the first century. He features in some of the most famous Icelandic sagas from the Middle Ages and was nicknamed "the Red" from the color of his flowing hair and beard. Everyone who comes to work on the ice in Greenland knows the story.

It started when Erik's father Thorvald was banished from Norway for manslaughter. Young Erik was still only ten years old, not yet a teenager, and having left behind all his playmates in his country of birth, he found himself living in Hornstrandir, Iceland's northernmost peninsula, with his family. But like father, like son, as the saying goes; when he was fully grown, Erik also killed someone. This was in 982 CE. The Icelandic court sentenced Erik to a three-year exile, during which the number of fights, aggressions, and sentences against him multiplied, as did stories of his many feats. Even the long Arctic nights wouldn't be enough to recount them all. Of all the legends featuring the exiled Erik, let me tell you the one that explains the baptism of Greenland.

The name originated almost as a joke. When Erik was banished from Iceland and arrived at the island, he looked at the arid land before him and realized he'd need something special if he were to attract new colonies to the ice shelf. That's when he had an idea worthy of today's top marketing gurus: He called the island Grønland, a name that would travel, telling tales of tender fruits and fertile land to unwitting sailors. Every time I think of Erik, that's how I imagine him, smiling smugly at his invention.

Although no one really knows what actually happened at the end of the first century in the far north of the world, other historical sources have suggested that Greenland was originally called Gruntland, as a number of ancient maps show. In this case, the *green* in Greenland is actually an error of translation: *Grunt* means "land," so it's not "green land" but "the land of land." Regardless of how the name actually came into being, the idea that Greenland has a color (green? white?) appeals to a lot of people's imaginations: to scientists, explorers, and even everyday people like Samantha.

Deciding Greenland's true color is not a purely mental or historical exercise, or even just a game to get through the sleepless nights on the ice. And it's not just an aesthetic issue, either. The color of Greenland is actually one of crucial importance for our much-loved planet Earth. Yet again, the reason has to do with the sun: The materials covering Earth's surface partially absorb (or reflect) solar energy by differing degrees. Fresh snow, as I mentioned earlier, reflects up to 90 percent of the energy coming from the sun, which is why it's so dazzling. By extension, it absorbs very little.

This also means that fresh snow melts much more slowly than older snow, because melting occurs when the snow has absorbed enough energy to transform water from its solid state (ice) to a liquid state. The same reasoning holds for the leaves of a tree, which, being darker than snow and reflecting more green light (why we see them as green), heat up much more quickly. Imagine being in the middle of the desert and having to choose whether to wear a white T-shirt or a black one: Which one would be best if you don't want to overheat? The answer is obvious. You don't need me to tell you.

The snow around me has a yellowish pall to it but, for once, it's not the impact of global warming, just my sunglasses. They were made especially for this mission, and along with the graduated lenses, which protect my eyes from direct sunlight, they have a soft spongy layer around the frames that perfectly fits my face. More importantly, they also block the light bouncing off the snow in all directions that, as I mentioned earlier, would be just as blinding as direct sunlight. When I wear them, it feels like I'm going into one of the tunnels on an Italian motorway that wraps you in a surreal orange light.

Different surfaces on Earth all reflect solar energy in a different way, each with their own spectral fingerprint of sorts. This stands for a property that varies along the electromagnetic spectrum, depending on the wavelength of the light. Objects that we see as red, for example, reflect more red light, while those that look white, such as snow, reflect light in several wavelengths, which combine to make the snow look white to our eyes.

An object's spectral fingerprint changes during its lifetime, a little like the way our hair color changes as we age, or the rings in a tree trunk increase or change depending on the phenology or age of the plant. The same happens with snow and ice: Freshly fallen snow can be distinguished from older snow by "reading" its spectral fingerprint. We collect basically the same data the satellites capture above our heads, both of which tell us something about the condition of the snow and ice through their spectral fingerprints—namely, the unique curve that every material in the universe has.

The graph obtained from our data shows a point for each color—each wavelength—by way of a digital alchemy that transforms the interaction of photons within the snow into a visible scientific language. Once we've downloaded the data to our computers, we identify which fingerprints (and therefore which snow conditions) match those in the satellite data and use this to build a map of the conditions of the snow and ice across the immense Greenlandic ice sheet. Moreover, our ground data will also be used to validate the satellite data. These special "cameras" in the sky capture images from heights of around 500 miles (800 km) while traveling at speeds of more than 12,400 miles per hour (20,000 km/h). The images are then sent to ground stations, where they are checked and uploaded to the web and made freely available to scientists.

In an article we published some time ago, we showed how Greenland entered a new phase in 1996. Since then, its albedo—the ratio of reflected irradiance to incident irradiance—has been decreasing. Ice alone absorbs up to 70 to 80 percent of the sun's energy, which means that as the ice below the snowpack gradually comes to light, Greenland absorbs more and more solar radiation and melts more as a result. I call this phenomenon "melting cannibalism." Ice is darker than snow, and its gradual uncovering triggers a chain reaction that accelerates melting: The more snow cover that melts on top, the more the icy layer below is exposed and the faster it melts. The dark, jagged ice that is left bare and wide open to the sun's energy melts much quicker because of the dual effect of rising temperatures and the greater amount of sunlight absorbed. Melting therefore increases progressively, a process that has been intensifying in recent years. Higher temperatures earlier in the spring season have preceded melting by several weeks and caused the blanket of snow to vanish earlier than usual. This exposes the ice below it earlier than usual, driving further and faster melting than in previous years.

And, finally, there's another factor influencing the "color" of Greenland, and that's the presence of fine dust and soot, blown by the wind, rain, and snow, and landing on the ice sheet. These particles come from many different sources: from sand from the Gobi Desert to ash from distant volcanic eruptions, fires in northern Siberia, and meteorite debris. When snow melts, these particles are partly flushed away with the meltwater and partly trapped on the surface of the ice; the more the snow melts, the more these particles build up on the surface, further darkening the ice. The effect is similar to when bare ice is exposed: The gradual acceleration of melting drives an increase in the number of dark particles released onto the surface, which in turn—in a sort of vicious cycle—further accelerates melting. The dark particles absorb more solar radiation than snowflakes do. It's yet another example of a glacial chain reaction.

To go back to the point that was puzzling Christine and Alison earlier, the secret is in the snowflakes, which grow in grain size with each cycle of melting and refreezing. In fact, when the snow starts to melt, the water pooling on the surface behaves like glue, fusing the snowflakes together in the thin layer of solid water that forms when the liquid water refreezes. This process affects how much sunlight the snow absorbs: The bigger the grains, the more solar energy they absorb. So yet again, we have another chain reaction, but this time a little different: To our eyes, the snow may look like it hasn't changed and is just the same as before, but it is actually absorbing more solar energy. This is because the change occurs at a different wavelength of light, what is known technically as near-infrared—outside the range of colors that are visible to the human eye. "Don't believe everything your brain tells you."

We advance into the light and toward our task for the day, which is to collect data. Armed with a special instrument called a spectrometer, a sort of artificial extension of our senses that enables us to see things that would otherwise be invisible, we walk toward the study area. As I walk across the snow, my mind goes back again to Samantha and her question about the color of Greenland. We might think of the country as an immense, desolate stretch of pristine, blinding-white snow, but the presence of dark particles on the surface, the larger grain size of the snowflakes, and the exposure of bare ice make it "darker" even when nothing seems to have changed to our eyes.

In the summer, when the Arctic sun shines constantly, radiating the ice sheet 24/7, the solar energy combines with the darkening of Greenland to accelerate the melting of the ice that much further. Since the phenomena we've been looking at are all amplified by the accelerated melting of the ice, the conclusion is fairly obvious: If, as scientists predict, Arctic temperatures continue to rise in future decades, the ice sheet will melt faster and faster.

Temperatures are rising here at twice the speed of the rest of the planet, turning what used to be Earth's refrigerator into a huge sponge leaking water into the ocean because it has reached saturation point. In the Arctic, even just a few degrees can make an enormous difference. A single degree might seem negligible when you're in the desert or city center, but here it has the power to change the entire system: from frozen and solid to melted and liquid. The melting of Greenland's glaciers, the ice sheet, and sea ice alters not only the local flora and fauna but also the entire ecosystem of our oceans. I reach our data collection point thinking about how precious the ice is. The stakes are high: Our planet itself is at risk.

CHAPTER 4

Forgotten Heroes

"IT'S NOT FAIR, it's not fair," I mumble over and over when Ian wakes me up, literally, from a daydream I've fallen into briefly. I wasn't sleeping, just dreaming on my feet. We get so tired at this latitude and after strenuous activity, I could easily drift off into a daydream the minute we stop.

I was thinking about when I left Italy. I was in Florence, at a bus stop near Santa Maria Novella, on my way to the headquarters of the National Research Council (I'd just completed my PhD), and I found myself chatting with a distinguished looking man I'd just met. We talked about this and that and I mentioned I'd be moving to the US in a few days to start a new job. From my accent, the man knew right away I came from the south of Italy, and with a look I'll never forget, part feigned compassion, part condescension, he asked me snootily, "Oh, so you're off to seek your fortune in America—pick up work as a shop boy or something?" My first reaction was to feel sorry for the guy, stuck in the past and still clinging to the outdated idea

of southern Italian emigrants "forced" to abandon their
homeland to find any kind of work. But that initial feeling was quickly followed by a mixture of anger and frustration. I was furious but decided to play him at his own
game; I complimented him for his discernment and ability
to read people so well, then bid him farewell after asking
him for a business card. Two weeks later I emailed him
from my NASA email address to explain what it is I actually do and invite him never to make assumptions about
people ever again.

Perhaps I overreacted, but things were different then.
I was always on the warpath, always ready to defend the
new life I was slowly building. You could say I was extremely touchy about anything that might offend my
origins or my past history in any way. I've been an "emigrant" for a large part of my life. First there was Naples,
where I studied electronic engineering and where I was
labeled the "Irpinian emigrant" despite being fairly close
to home. (It might sound crazy but it's not really when
you think about it: Neapolitans and Irpinians are two different strains, even though the distance between the two
cities is less than that between my home in New York and
JFK airport.) Then there was the time in Florence when
"Irpinian" became synonymous with "southerner." And
finally, in the US, where integration (racial, gender, etc.)
has evolved more than in Italy, although pockets of con-
servative America still particularly indulge in social and
racial discrimination.

I reassure Ian and tell him everything's fine. In a few words, I tell him about my daydream. He looks at me for a second, confused, then retreats into a respectful, considerate silence. He falls in beside me as we walk, just him and me, the others behind us. We're roughly the same age—both with children verging on adolescence. We seem to understand each other. At times like this I feel I can share thoughts with him that come as a surprise even to me: feelings that emerge from the laying bare of the soul, like nunataks, the rocky peaks once buried in the ice that pierce the pristine white mantle around us. Ian casts an eye toward the others. They're close enough for us to be part of a group but far enough not to hear what Ian and I are saying. We talk some more. In Greenland, I think more often about my daughters, young women in a world ruled by men. I wonder how to be there for them, help them achieve their dreams, give them the courage they need to get ahead in life. I also think about Anitra, my partner. An American born to a Mexican mother and African American father, she often tells me of the racial and social prejudices she has endured.

What is courage at the end of the day? How do you measure it? What value scale do you use? My mind turns to the hidden, lesser-known part of the recorded history of polar exploration, led not by the famous Nordic explorers we've all heard about, but spearheaded by individuals who remained far from the spotlight—the minorities who were often overlooked in coverage of the great expeditions they were part of.

Matthew Alexander Henson is one of these and, without doubt, one of the most fascinating explorers to have ventured into the Arctic. He was born a century and a half ago (August 8, 1866) in Nanjemoy, Maryland, to freeborn African American sharecropper parents. Like many other African Americans after the war, Henson's parents experienced frequent attacks by the Ku Klux Klan and other white supremacist groups. To escape the racial violence, the three-year-old future explorer and his family moved to Georgetown—a small, unincorporated community in Maryland that today is now a fashionable residential district of Washington, DC.

Henson's mother died when he was young, and when his father also passed away not long after, he went to live with his uncle in Washington. The uncle cared for Henson and paid for his education, but unfortunately the young boy seemed to be under some sort of curse, as only a few years later his uncle also died and, at age eleven, he was forced to look after himself.

It was during this period that something important happened, something that would change the course of Henson's life: He heard a speech. When he was ten years old, he attended a ceremony in honor of Abraham Lincoln, where he heard the renowned orator and activist Frederick Douglass speak. A leading figure in the black community, he called upon all those present to take advantage of any and all educational opportunity that came their way and to remain steadfast in the fight against racial prejudice.

When he turned twelve, the young Henson headed off to Baltimore, Maryland. It was in the busy port that a new life opened up for him as cabin boy aboard the merchant ship *Katie Hines*. He sailed for several years on a journey that took him to distant, unimaginable places such as China, Japan, Africa, and, most importantly, the Arctic seas. It was during the latter part of his time at sea that the ship's leader, Captain Childs, decided to teach Henson how to read and write. But in 1884, when Henson was twenty-one, Childs died, and Henson returned to Washington, where he found work as a salesclerk in a clothing store.

It was here, in 1887, that he met the explorer Robert Edwin Peary, who was so impressed with Henson's seamanship that he took him on as his personal valet for a planned voyage to Nicaragua. After returning from Nicaragua, Peary found Henson work in Philadelphia. In April 1891, he married Eva Flint. But the call of the sea was too strong, and it wasn't long before Henson joined Peary on another ambitious expedition to Greenland.

This voyage was a revelation for Henson. He embraced the local Inuit culture, learning their language and their survival techniques. In 1893, Henson returned once more to Greenland, this time to chart the entire ice cap. It was a long and complicated voyage that lasted roughly two years and ended with Peary's team forced to eat all but one of their sled dogs.

Despite this perilous trip, the explorers returned to Greenland in 1896 and 1897, their will and perseverance as strong as ever. Their mission this time was to collect three large meteorites they had found during their earlier quests, including one that weighed around 34 tons (31 metric tons), the largest ever found—and which were sold to the American Museum of Natural History and the proceeds used to help fund their future expeditions. (Don't forget, as I mentioned earlier, how Peary got his hands on them!)

Henson's frequent absences took a toll on his marriage, and in 1897, he and Eva divorced. Five years later, in 1902, he mounted his first attempt to reach the North Pole by Peary's side. The attempt ended, as is well documented, with the tragic death of six Inuit members of the expedition. Their next attempt in 1905 was backed by President Theodore Roosevelt, and the group managed to sail within about 300 miles (500 km) of the North Pole. Thick ice thwarted their passage—the same ice now melting alarmingly fast, which now allows for a much smoother sea passage northwest—and they were forced to turn back. Around this time, Henson fathered a son with an Inuit woman and named him Anauakaq. But when he returned home in 1906, he married a woman named Lucy Ross.

The team's final attempt to reach the North Pole began in 1908. Henson proved an invaluable team member, building sleds and training others, who considered him to be Peary's equal in terms of Arctic experience, on their handling. On April 6, 1909, the forty-three-year-old Henson, along with Peary, 4 Inuits, and 40 dogs, finally reached the North Pole—or at least they claimed to have reached it. What the feat had cost them was clear to everyone: The expedition had started out with 24 men, 19 sleds, and 133 dogs! Peary knew that the mission's success was chiefly down to his trusty companion's skills, and he stated publicly that he would never have made it without him. If truth be told, the success of Peary and Henson's expedition is still widely debated. Some believe they only thought they'd reached the North Pole but had actually miscalculated their route; others are convinced Peary intentionally lied. Frederick Albert Cook also claimed to have reached the North Pole a year before them.

On their return, Peary received many accolades, while Henson—due to the color of his skin—was largely overlooked and spent the next three decades working as a clerk in a New York federal customs house. But he never forgot his life as an explorer. His Arctic memoir, *A Negro Explorer at the North Pole* (later retitled *A Black Explorer at the North Pole*), was originally published in 1912, but he didn't receive the acknowledgment he deserved until he was seventy years old. In 1937, he was accepted as a member—the first ever African American one—of the Explorers Club in New York, and in 1944, Congress awarded him and other members of the expedition the Peary Polar Expedition Medal.

Ian and I both have teenage children and often find ourselves discussing what the future holds for them and how society—and our profession in particular, despite the important progress that has been made—is not the easiest for women to gain access to. Anything but. Our conversations often turn to the role women have played in Arctic and Antarctic exploration and how hard it must have been for the first trailblazers to gain any ground. Many of the first women involved in exploration of Antarctica were, unsurprisingly, the wives of explorers. Those who wished to have larger roles in missions to the great icy continent had to overcome sexist ideas and surmount bureaucratic inertia. Because of the pervasive patriarchal culture, women who were qualified for expeditions in Antarctica were less likely to be selected than men.

In the early days of Antarctic exploration, men often thought of the continent as a place where they could imagine themselves heroic conquerors, and in their diaries they referred to the great white mass as a "virginal woman" or "monstrous feminine body" to be conquered by strong and virile men. Women were often called upon when it came to the naming of new territories that had been discovered and were even encouraged to have babies in Antarctica. It might seem bizarre now, but thirty or forty years ago, several governments invited women to give birth there and stake a sort of national claim to the area, given that no country officially owns the continent. States are merely granted access to "use" the Antarctic under a treaty of international cooperation. One of the first women to accept the invitation was Argentina's Silvia Morella de Palma, who delivered 7-pound-8-ounce (3.4 kg) Emilio Palma at the Argentine Esperanza base on January 7, 1978.

Ian remembers an article from the mid-'90s that, complete with statistics, showed how women can better withstand the extreme cold of the Antarctic. I file this piece of information away until such times as I can share it with my daughters. We continue our conversation, remembering the heroines of Antarctic exploration. The first European woman to reach the waters of the southernmost continent was Louise Seguin, on the *Roland*, in 1773, although it's not clear if she joined the ship as a courtesan or disguised as a cabin boy. Others believe the first woman was Jeanne Baret, a French botanist and explorer who traveled to the sub-Antarctic region for scientific research purposes and became, without knowing it, the first woman to circumnavigate the globe.

The first woman in the modern era to set foot on Antarctica was Caroline Mikkelsen, in 1935. Or so the world has always thought. Born in Denmark, Caroline later moved to Norway when she married Captain Klarius Mikkelsen, whom she accompanied on a resupply mission he was leading to Antarctica. It wasn't clear at the time if Caroline had actually landed on the Antarctic mainland or an island. It was only early in the new millennium, after her death in 1998, that researchers published articles concluding that she hadn't, in fact, reached the mainland. She had actually landed on an island a few miles from the coast. I can only imagine how frustrated she would've been, if she'd known how close she had gotten—to have missed the target by such a tiny distance.

Let's look at the numbers: The distance between Antarctica and Europe is on the order of some 6,000 miles (10,000 km). The 3 miles (5 km) that Caroline missed the mainland continent by equal 0.033 percent of the total distance she traveled between Europe and Antarctica. Missing the target by 3 miles (5 km) is a little like saving for years and years to pay for a $1,000 holiday and having it fall through because you're just $0.33 short. Nevertheless, it resulted in the title of first woman to set foot on the Antarctic mainland being conferred upon Ingrid Christensen instead. The Norwegian daughter of one of the most successful whaling magnates of the period, Ingrid was a role model and natural leader for women of her time: She was fearless, charismatic, and incredibly independent.

In 1936–37, Ingrid made her fourth and final trip south, accompanied by three other women, one of whom was her daughter Augusta Sofie Christensen. Ingrid flew over the mainland, becoming the first woman to see Antarctica from the air. On January 30, 1937, Lars Christensen, Ingrid's husband, recorded in his diary that his wife landed at Scullin Monolith, a crescent-shaped rock fronting the Antarctic Ocean, becoming the first woman to set foot on the Antarctic mainland.

The others have pulled closer to us and Ian heads over to them. I remain to one side, preferring to be alone a bit longer. It helps me to focus my thoughts on the tasks ahead. I go over in my head all the things not to do, the mistakes not to make, the little things that could ruin everything. Then my mind drifts back to women in Antarctica and my daughters, and the many times I've dreamed of sharing these emotions and experiences with them, now and in the future. Even though they're still too young, I imagine myself chatting with them right here, in this exact spot.

My mind jumps, by association, to the letter I wrote to them recently, when I was working at the edge of the ice sheet. I take the letter everywhere with me. To be fair, it's actually a copy, so I don't ruin the original, which I wrote on a sheet of paper ripped out of my scientific notebooks. I read it before I set off to join the others and it infuses me with new strength and a warm feeling of comfort.

Dear Olivia and Francesca,

Looking at the shimmering, imposing ice here in Greenland makes me feel so at peace and in harmony with the world that I want to write these few words to you. Everything has been so much richer, so marvelous, since you were both born. One day, I hope we'll be able to talk about why we have to live apart. I'd like to explain the reason why. But there's no rush, I'm sure life will give us the opportunity. It's hard for me to live, day after day, without the sound of your presence, your noise, and your color. I imagine it's even harder for you. After all, none of this was your choice. But despite that, both of you—young explorers of your future among the conglomerate of life in our universe— you've never held it against me, never complained. I hope these few words can convey how proud I am of you and how lucky I feel to share my life with you, to be your father. Not for one second have I ever felt judged by you. If anything, you've been the ones to make me feel your presence, your support, your love. All this gives me the strength to try and be a better person, a better father— respecting the decisions you make the same way you have respected mine.

Be free! Find joy in your life, nourish love, friendships. Tend the beauty within and around you. But don't allow your quest for freedom to morph into egotism: Find a way to feel loved while loving back with respect. Unlike what some people say, I believe freedom and love can walk together in the journey that is a relationship. In truth, it's the only way to reach the summit of true love, on a pathway that must be discovered day by day, hour by hour, emotion by emotion. Fill your souls with knowledge and emotions. Grow in every direction; explore inside and outside of yourselves; invent tools that will help you to dig where no one has ever dug; respect the choices of others, but don't be afraid to voice your opinions; embrace life as if you can't change it. Trust that time will bring change. After all, you are time.

The noise of the water gurgling in time to these words reminds me of the cycle of life. I think about your lives, the future that lies ahead of you, and how privileged I feel to witness the flow of your existence, which, like this river, whispers continuously to my soul and cradles my heart when I think of you. My love for you is rooted in ice so thick no one could ever melt it.

CHAPTER 5

Arctic Big Brother

A FEW HOURS LATER, we arrive at our first data collection spot and remove our harnesses and backpacks. We won't stay in the same spot for long, though. The plan is to move with the spectrometer from one point to another, stopping along the way to capture the spectral fingerprint of the ice we're walking on. The spectrometer measures with extreme precision the amount of light reflected by the snow at every wavelength. It's a technological marvel: Light is directed along an optic fiber, bounces off a system of mirrors and gratings inside the instrument, and is broken or "split" into its component colors (or wavelengths). This data is sent over a Bluetooth connection to a laptop near the instrument. The spectrometer itself is kept inside a small special backpack that's compact but heavy. The meter-long cable running out of the bottom of the backpack connects to a long rod of the same length. Light enters the sensor at the end of the rod, at a right angle to it. It's not the easiest configuration but it means we can "have a stroll" and take measurements from multiple sites, extending the range of our ground data and giving us a better picture of what's happening here in Greenland.

There's a specific procedure to follow: The device has to be switched on several hours before we go out to collect data, to give it time to "warm up" and make sure the changes that occur in the electronic components when the power is switched on (because of the change in temperature) don't skew our data. You have to be very careful not to overflex the fiber optic cable and block the "light highways" inside it or break the cable itself. Then there's how to actually "wear" the instrument, which requires a second person, along with the "wearer," to mount it in position—a bit like astronauts gearing up for a space walk.

Patrick is our "astronaut." He points the sensor at the sky first, over his head, to measure the solar energy hitting the surface and using the digital spirit level on the rod to keep it in parallel with the ground. Then he presses the button to start capturing data, and a beeping noise, a bit like an incoming text message, confirms that the data has been logged. The sensor is then pointed toward the ground—at the ice, the rod held out, arm straight. He presses the button again, there's another beep, and the first full measurement is complete. We move on, a few feet further ahead, and do the whole thing again, repeating the steps for a couple of hours, trying to gauge how much of a surface area we can cover in the allotted time: Measurements have to be taken while the sun is at its highest point in the sky to assure their accuracy. In total, we have one to two hours at the most, starting from when we arrived. We need to be fast and accurate, two qualities that get increasingly challenging the colder and more tired we get.

It will be a long time before we know if the experiment was successful. We have to wait until we're back in New York and have the data processed in the lab. It's a side to the job that, understandably, might seem frustrating at times, but in some cases, to me anyway, it can make it exciting. It's a job that requires determination, perseverance, creativity, and initiative, not to mention nerves of steel and quick reflexes to make sure nothing happens that could jeopardize the many months—sometimes years— of meticulous preparation prior to the mission. All these qualities are required to find solutions to problems you've probably never had to deal with before and perhaps could never have imagined, using the few resources you have available on the ice.

I think about this every time I see these graphs in articles of my own or published with colleagues; every piece of data, each one of those tiny points, belie a monumental amount of legwork—months of preparations, years of hypothesizing and planning, endless conversations, discussions, thinking, and reflecting, out loud and in our heads. All this to get to a minuscule dot on a graph, a bit like the contraction phase of the universe: An idea explodes, is tested, then collapses to a dot—one dot that joins lots of other dots to map our numerical passage through the country we traversed and studied.

I remember a time—it was June, I was in the center of Greenland, at an altitude of about 6,000 feet (2,000 m) caught in a surprise snowstorm—when our steam drill, a device that drills cylindrical holes into the ice (to hold the aluminum poles that support our equipment), gave a deafening boom and stopped. Steam escaped from a nozzle at the end of a pole a few feet long and the gas tube had small cracks in it. We managed to get it running again thanks to some insulating tape, a collection of nuts and bolts I happened to have in my pocket (when you're on a mission in a remote location like this, you never know what you might find in your pocket), and a ball of twine one of the party had brought with him to wind around his fingers to relax. Who would've thought? A million-dollar investment saved by a dollar-store fidget toy.

Once we're done with the measurements, we stop to eat. Finally. We've been out since early morning and hunger has set in. When I'm out here on the Greenland ice, I'm always hungrier than usual. Then again, I've always had a fairly fast metabolism and am used to eating a little and often. But out here I need a lot more energy than usual. Because of my body type—thin but not frail—I need a lot of fuel to keep warm. And also to fuel my brain, which tends to be hyperactive on these missions, burning an endless supply of energy.

It's lunchtime—if I can call it that, given we can't afford the luxury of a long break, or even to sit down around a table to savor something tasty. We use the humps of ice around us as stools, the surface of the ice sheet as a table, laid with a multipurpose mat. We made up our lunch this morning and it is composed of hot soup, a sandwich, and a bar of chocolate. Oh, and coffee, of course—by the gallon. We boiled some water at breakfast and carried it in flasks to keep it warm; now we pull out the packets of dehydrated soup, which have different names and supposedly come in different flavors, although, to be honest, they all taste pretty much the same.

Nevertheless, the hot soup and the mineral salts it contains, as well as the intimacy of the moment out here together against such a stunning backdrop, combine to regenerate and replenish the physical and emotional energy we need to carry on with our tasks. Mealtimes are something of a ritual. You need to act swiftly when you're emptying the packet into the water, otherwise the water will get cold and you end up with soup that's not only cold but also raw and lumpy. In other words, a *ciofeca* (like crap!), to coin a popular phrase by Totò, my favorite comedic actor. We put the lids back on the airtight containers and wait for the soup to be ready. In the meantime, I talk to the others about how hard it is to collect ground data, and someone mentions that huge steps forward have nevertheless been made this sector.

I remember when I was doing my PhD in the late '90s, terms like *petabyte* and *gigabyte* were the exclusive domain of a handful of experts. A petabyte is one million gigabytes, or more than one quadrillion bytes, and a byte, as you probably know, is the unit of measurement used for computer memory or the size of a data archive. To give you an idea of the relative dimensions, a text message (SMS) from a mobile phone equals 140 bytes, so a petabyte would be eight trillion texts. With today's technology (due to be even quicker in a few years' time), we can download data to our laptops at speeds that were once unimaginable. My early experiences at NASA, where we'd have to sort through mountains of floppy disks sent to us in enormous boxes, then spend endless sleepless nights uploading them, one by one, to our computers, now seems like something out of a prehistoric research project. The explosion of the web and multimedia technologies changed everything: Data is basically generated, shared, and viewed in real time. The so-called Big Data Revolution that influenced and transformed every area of our lives via smartphones, laptops, smart TVs, and so on and so forth, has also, and perhaps most importantly, changed the way we do scientific research.

Lunch should be ready. We open our flasks and tear into the hot soup. It's not as piping hot as we expected, but the heat of the first sip is instantly invigorating. It relaxes us, and we chat. Some dip their cheese-and-butter sandwich into their soup, others swallow it down quickly to avoid burning their lips.

The conversation about data continues. Researchers at the University of California tried to quantify how much information there is in the world and how to convert this figure into something tangible that we humans can comprehend. Let's look at some numbers: Around ten zettabytes of data are processed every year, which means one quintillion gigabytes. Mind-boggling, huh? To give you an idea of just how big this is, there are "only" about 250 billion stars in our galaxy, the Milky Way, which means we'd need another four billion galaxies like ours to equal the number of bytes of data processed every year. The Californian researchers also calculated that if we were to work out the storage capacity required for all the mobile phones, laptops, and computers needed to contain the tens of billions of billions of bytes generated every year, we'd need a never-ending flotilla of 800 aircraft carriers for the mobile phones, 250,000 planes for the laptops, and the equivalent of 900,000 Statues of Liberty for the desktop computers. The quantity of data we gather here in Greenland is nothing compared to this: just a few gigabytes. It's when you sum it all up that it gets out of hand—and, as the years pass, the figures will just keep growing.

It will come as no surprise that one type of information that is already growing exponentially is from satellites, which are like technological fortresses—knights of sophisticated armor overseeing our planet and vital allies in the race to understand how it works. Satellites orbit hundreds of kilometers above our heads at speeds of around 18,000 mph (30,000 km/h—more than twenty times the speed of sound), observing everything, perceiving things that are invisible to our eyes as they elicit the secrets of clouds, snow, ice, trees, and everything else that lies below them.

According to estimates, around five thousand launched satellites are currently orbiting in space. Of these, only 40 percent are still operational, meaning that there are around three thousand useless pieces of metal circling Earth—little more than space debris. It might sound funny at first, but the number of "dead" satellites is actually a major issue, partly because they could drop to the ground at some point, crushing both homes and humans, and partly because satellite pollution interferes with the launch of new satellites, not to mention the nuisance factor for the people working on space stations and performing zero-gravity space walks.

The majority of the civil satellites orbiting our planet (very little is known about the military ones, such as how many there are and what purpose they serve) are used for telecommunications, primarily mobile phones, internet, and GPS, the system we use every day to locate our position and orient ourselves whether we're on foot or in the car. Lucky for us, a large number of satellites have been assigned to study the planet with all nature of technology, from high-resolution digital cameras to microwaves.

The earliest remote-sensing experiments date back two hundred years and—according to a very specific definition—were performed for the purpose of acquiring information about an object or phenomenon without making physical contact with the object. These attempts were being made at the same time as two major revolutions of the modern era: photography and the development of flight. In the late nineteenth century, in fact, the first pioneers began to take landscape photographs using the first primitive cameras, looking down from a marvel of technology at the time—the hot air balloon. The first ever aerial photograph (which probably no longer exists), in 1858, was taken by Gaspard-Félix Tournachon, a Frenchman, and it was a view over Paris. The oldest surviving aerial photograph, though, is one of Boston taken from a tethered balloon by the American photographer James Wallace Black in October 1860.

After this it was the turn of the pigeons, which were used during the First World War to take photos of enemy territory without being seen or risking human life. But it was with the advent of modern aviation, the development of technologies to launch missiles into space, and technological advances made in photographic equipment that the real "boom" came in remote sensing. From the spy planes whose photos triggered the thirteen-day confrontation between the US and Cuba to minicameras on drones, this technique has continued to evolve until the present day and is now one of the most valuable tools we scientists have, on a par with satellite or aerial observations.

The first satellites were launched in the 1960s and were used for climate monitoring and to improve weather forecasts. What we tend to forget is that a large part of the progress made—both technological and scientific—in studies of our planet has come from the defense industry and its technologies. Radar, for example, an instrument we're all familiar with, used in both civil aviation and on Earth-monitoring satellites, was invented and perfected primarily for the purpose of fighting wars. This is a controversial issue in the scientific and research community and has always created, is creating, and will continue to create heated debate. On the one hand, scientists owe a lot to the defense sector for the financial and technological boost it gave the remote-sensing sector; it afforded us greater insight into how our planet is changing. On the other hand, most of the people I work with (and I include myself in this) are fervent anti-militarists.

The soup, or what's left of it, is cold by now. My fascination with stories and looking at things in depth has got me in trouble yet again. It's time to move on to dessert: For want of a better word, "delicious" protein bars are savored as the bowls and cutlery used for the soup are put back in our packs, careful not to leave anything that could contaminate this spectacular setting.

The conversation goes back to satellites. The data we've gathered today will help us to better interpret the satellite imagery and understand to what extent and in what way Greenland is changing. From a professional perspective, we're lucky to live in the times we do, benefitting from the staggering rise in the number of Earth observation satellites in recent decades and the vast improvements in the quality of data received. Many areas of the North and South Poles are now "covered"—as we say in the business—by daily satellite tracking, with northern and southernmost points observed several times a day.

Ian is an expert on modeling but not satellites, so he asks me what the most "popular" ones are these days. Before I reply, I take a sip of coffee to warm me up and wet my lips, which have dried with the cold and the chat. It would be impossible to name them all, obviously, but the most famous ones are Landsat and MODIS, then a handful with alien-sounding acronyms—SSM/I, AMSR-E, SMMR—and other more self-explanatory and associative ones like Terra and Aqua or ICESat and CloudSat.

I cobble together a general overview, although whatever I say could never do the subject justice: There's just so much to know that there are even specialist degrees in remote sensing these days, and we only have a few minutes as it's almost time to get going again. To sum up, there are two types of sensors: active, which emit pulses of electromagnetic waves and measure how many come back, and passive, which listen in ("passively") to what's going on around them. An everyday example of the active type would be a flashlight: The beam of light falls on an object and our eyes see its color, shape, and size, which enables us to identify it. In this case, the flashlight is the electromagnetic source (of light in the visible spectrum) and our eyes are the sensor. To imagine a passive instrument, think of the natural energy emitted by Earth: In physics, every object in the universe transfers energy through electromagnetic waves; remote sensing with passive sensors is a little like listening to someone's electromagnetic breathing.

There are advantages and disadvantages to each technique: Digital cameras (classified as passive sensors because they detect naturally emitted light from the sun and not a source emitted by the camera itself) can reveal interesting details, but there are clearly limitations—like the presence of clouds, for example. Microwave sensors, on the other hand, can "see" through the clouds, but the images produced are not as clean. In other words, we have to accept a compromise: high-resolution images that are not always available due to cloud cover, or lower-quality, "hazy" images lacking in detail but always available.

Patrick snaps out of his post-lunch lethargy to remind me that not all satellites use electromagnetic waves, and that gravity sensors are also used. The GRACE satellite (Gravity Recovery and Climate Experiment) detects changes in Earth's gravity field, literally "weighing" our planet from space. It does this using twin satellites that take precise measurements. As they rotate around our planet up to fifteen times a day at an average speed of around 1,200 mph (2,000 km/h), their exact position is influenced by Earth's gravitational pull, which varies as mass on Earth varies from place to place. So, by way of example, when the first satellite passes over an area with a stronger gravitational pull, it accelerates, and the distance between it and the other satellite increases. The wonders of human ingenuity. Not even Isaac Newton, who formulated the law of universal gravitation, could've imagined how gravity would be used in the not-too-distant future to study our planet using an orbiting satellite. The system is so precise that the distance between the two satellites can be measured to an accuracy of one tenth of the width of a hair. The data gives a clear idea, for example, of the mass lost in summer by Greenland or Antarctica and in turn allows us to calculate—with precision that was previously impossible—how much the two immense ice caps are contributing to rising sea levels.

We go back to our measurements, hoping that they turn out OK.

Edge of ice sheet in Kangerlussuaq
(Greenland)

Greenland's ice sheet, photographed
from the helicopter, prior to landing

Icebergs near Ilulissat
(southern Greenland)

McMurdo Dry Valleys
(Antarctica)

Glacial meltwater
stream (Greenland)

Canyon carved in ice
by water (Greenland)

Iceberg in the bay at Ilulissat (Greenland),
photographed during a midnight boat trip

View of the Greenlandic
ice cap from the helicopter

Hole in the ice generated by water draining from
one of the lakes on the surface (Greenland)

Our base camp near one
of the lakes in Greenland

View from the tent before
going to sleep (Greenland)

CHAPTER 6

Icy Abysses

LUNCH HAS BOOSTED OUR ENERGY and enthusiasm levels. We needed it. You won't find it written anywhere in ice expedition manuals, but staying motivated is key to the success of any experiment. Everything seems different when we set off again: We're walking faster, the backpacks don't seem as heavy, the shoulder pains have abated (they'll be back soon, we know—possibly more painful than before). I feel more supple, am moving more freely over the ice—ever cautious though—using a technique we've more or less perfected now. It comes fairly naturally and entails harnessing the "momentum" of each step to expend less energy on the next one, a bit like the hopscotch game we used to play as children. You find a stone and use it to draw some squares on the ground, throw the stone along them, then hop and jump, staying inside the lines, to the square the stone landed in.

We head to the site of our second experiment. There are no signs of crevasses, but you can't afford to let your guard down on the ice: Even a simple ankle sprain, if you get your foot caught in one of the many tiny fissures across the surface of the ice, or cutting your hand on a jagged ice wall can be a serious issue in a place like this. Not to mention more mundane things like chatting to an expedition companion and not noticing that the rope we're all attached to has become tangled around someone's leg or wedged on a hummock. These may all seem like minor distractions, but they can be fatal nonetheless, for individuals and for the expedition as a whole.

A quick glance up at the horizon and down to check our footing, we see the first signs that we're nearly at our next site: The deep thrum of water, maybe a distant waterfall, reminds us of how accustomed we've grown to living in the absence of sound. The rivers and creeks around us slowly increase in number: We can't be far now. We emerge from the final gully and crest the small hill between us and the lake we've come to study. We christened it "Lago Napoli." It was my idea, partly to celebrate my roots and partly as a bit of fun, to play our Swiss colleagues in a game they started, two decades ago, when they named one of Greenland's first permanent observation posts the Swiss Camp.

In satellite photos, the lakes we're studying look like glimmering gemstones bejeweling the ice. They're formed from the purest glacial meltwater, which pools in depressions on the ice surface; looking at them, you can see all the way to the icy bottom—the water is so crystal clear. I remember vividly the day I saw them for the first time from a helicopter: It was like white giants—imaginary Gullivers in the white Lilliputian expanse—had let droplets fall to the ground from high above.

In scientific terms, these lakes are an object of study for a variety of reasons. Being darker in color than the surrounding ice, they absorb more of the sun's energy, which has a warming (melting) effect on the ice. One thing that really fascinates and floors me at the same time is how alive these lakes are: The bottom is not static; they're not immobile things, fixed like their mainland counterparts. Instead, they're constantly evolving, changing, growing deeper and deeper as the life cycle of the lake progresses. This isn't just a poetic way of evoking the stark purity of our surroundings; it's a scientifically proven fact. On a past expedition, we measured the bottom of these lakes and saw that they reach depths of around 6½ feet (2 m) in a couple of weeks. It's like they feed on themselves, slowly gouging the ice with a simple yet effective action that wears it down. In other words, they're "cannibal" lakes.

Another key but not quite as obvious feature of these shimmering pools is their potential connection to rising sea levels. The presence of lakes on the glacial surface triggers a process similar to fracking called hydrofracturing: Ice flowing toward the sea causes the surface to fracture under its own weight, like lava creeping down from the summit of a volcano, creating crevasses and fissures along the way. The ice on the top moves faster than the ice sliding along the bedrock, and the vertical pressure this generates forces open fissures in the ice; these cracks normally close as part of the natural "concertina" movement of the surrounding ice, accelerating and slowing. What changes this is when there are lakes on the surface of the ice. The water gushes into the crevasses, exerting pressure on the walls, which not only keeps the cracks open, but pries them further apart. Basically, water wins over ice. But to maintain its upper hand, water needs to keep up a constant pressure to widen the crack.

This fact in itself is fascinating for researchers like me. It's like witnessing the perfect battle between two physical states of matter: liquid and solid. There's water, the raison d'être of our planet, in the liquid state, molecules jumping around, vibrating convulsively (the so-called Brownian motion), chaos reigning supreme. And then there's the immobile solid state—crystalline structure, composed beauty, and rigid elegance. Lakes have won the battle here, their glacial waters driving open crevasse after crevasse, relentlessly prying them apart, widening and deepening them, driving them down to the bedrock the ice itself rests on. When this happens, it's a bit like pulling the plug out of a bathtub full of water: The contents thunder with unrestrainable violence down through the newly opened vertical shaft.

Our plan is to lay sensors on the bottom of the lake to find out when and how fast the lake will disappear. We also want to lay more sensors around the lake to monitor the

movements of the ice before, during, and after the catastrophic draining. But we can't go into the lake on foot to
lay them. We will be the first to measure or film a glacial
draining using only the tools normally available to us. We
considered going out in a rubber dinghy, but the idea was
rejected immediately: The force of the water being sucked
down through the conduit in the ice would drag us with
it, and we wouldn't last more than a couple of minutes in
near-freezing water.

We hit on a solution a few months ago and it's finally
time to put it into practice: We're going to use a radio-
controlled boat sourced specially for this mission. It's
basically the kind of device fishermen would use to drop
bait as far as possible into fish-rich waters that they would
have no other way of reaching, only we'd modified ours a
little. Fishermen control the boats from the side with a
joystick, drop bait at the required spot with the press of
a button, and, when the fish bites, pull the boat straight
back. I must've looked at hundreds to find the right one
for our mission: Whenever I explained to manufacturers
what I needed, they would look at me like I was bonkers,
especially when they pushed for more details and I had to
explain that it was for a NASA experiment. We eventually
found a boat in Scotland and got to work installing a GPS,
seabed scanner, spectrometer, and customized devices we
created ourselves to mount this equipment onto the boat.

I have such vivid memories of the day we ran the final trials in New York, prior to shipping the boat out to

Greenland: It was a sunny spring day in Central Park and we'd decided to run our trials in one of the ponds. We tried to get a written permit but ended up just speaking to the park manager, who was happy to help and told us to go right ahead. The boat, it has to be said, attracted a bit of attention: It was black with colored LEDs on the front and back, technologically somewhere between a catamaran and the Batmobile. Everything went smoothly, the boat responded perfectly, and even drew a small audience of bemused tourists, until two plainclothes police officers arrived, that is. Unimpressed at our splashing around in the lake, they ordered us to leave.

We arrive at the edge of Lago Napoli. It's finally time. Backpacks are dumped on the ice, the instrumentation mounted onto the boat, the vessel powered up. We guide it carefully out onto the icy water. The video camera attached to the side sends us images of the lake bottom as it motors away, directed by light flicks of my fingers on the levers of the radio controls. At around 6 feet (2 m) in length, the boat looks like nothing more than a tiny, black dot when it's finally in position. I can't tell the bow from the stern anymore.

There's nothing more we can do to control it now. We might not be at sea, but the wind blowing here in Greenland is enough to ripple the surface of the lake with waves up to 1½ feet (½ m) high at the center—not huge, but high enough for a miniature boat. It's time to anchor it. We're all tense but trying not to show it, trying to stay calm. Before I press the button to release the sensor, I look around, searching for approval, as if it's a nuclear bomb I'm about to launch.

It's done! Months and months of preparations gone in a hundredth of a second, our fate now in the hands of the electrons carrying the charge, which turns into a radio wave on its way to the boat, hundreds of feet away from us. We can't see anything at all now—the boat is just too far away. The video camera was also mounted for this reason, so we could watch from our computers. Hoping that something resembling our sensor—a metal tube about 6 to 8 inches (15 to 20 cm) long and no more than 2 inches (5 cm) wide—will soon pop up on the screen, we huddle anxiously around the laptop. It's like watching penalties in a World Cup final. Finally there's a glimmer: Our sensor appears. Joy mixes with relief in an indescribable whirlwind of disjointed emotions, in the middle of which we suddenly realize how close we'd come, in that brief few minutes, to the success or failure of such a huge part of our expedition.

We pull the boat back: Everything seems in order when we take it out of the water. Our adrenaline levels are so high we're not even aware of the cold pressing in on our hands after the contact with the icy water and wind. This is not the end—we're all acutely aware of this—but, for now, all we can do is wait. Weeks could pass before we get the data we need. There's still no clear idea about what happens when these lakes suddenly collapse in on themselves. There are multiple factors at play: the amount of water in the lake itself; when the vertical shafts formed in the ice; how long they've been there; what's at the bottom; what the bedrock at the base of the ice sheet is made of; and if there are valleys and mountains buried under the weight of this colossal mass of hydrogen and oxygen.

Back at camp, we've just finished storing the equipment and instrumentation in our backpacks and boxes when Christine, who hasn't been able to take her eyes off the screen, asks me to look at the images coming once every minute from the camera at the edge of the lake. She wants to know if what she's seeing is real or some sort of optical illusion. Puzzled, I join her at the computer. At first glance, I don't see anything out of the ordinary—just the side of the lake. We watch the sequence of photos again together, scroll quickly forward and back, generating a flux of moving images that flash past so quickly it's like a digital merry-go-round. I stop the sequence: I can see it now. Christine's right. The edges of the lake are agitating more than they should and not because of the wind; the movement we're seeing is too broad to be wind, too synchronous, and reaching too far across the whole surface. What's more, reports from our weather station signal that the wind has actually died down. It dropped off rather suddenly, as can happen in this part of the world. And another thing: The water level is also dropping, albeit imperceptibly.

We look at each other, puzzled, scared to accept, perhaps, that what we're witnessing in real time is the event we came all the way here to see: The lake is collapsing, imploding on itself. Before we get the chance to shout for the others, an earsplitting boom captures the attention of every one of us, including those who'd gone most numb with the cold, the wait, and the fatigue. We're now excited, on edge, curious. Those who can grab their binoculars peer into the distance; the others content themselves with the computer screen. Every minute that passes feels like an eternity. We're glued to the screen, consumed by curiosity. Huge blocks of ice seem to be spinning on the surface of the lake as the water is swallowed up into something that looks like a giant bath with the plug pulled out. The fragments of ice must've been thrown out by the force of the event. They look featherlight as they glide through the air, but they have to weigh as much as a bus. We keep watching the slow agony of the lake, which will disappear in less than forty minutes, the water flowing out faster than Niagara Falls. The hole in the ice is too big and has no time to heal before the lake's lifeblood—the water—is drained away in a sort of glacial hemorrhage. We can't resist any longer, we have to see it with our own eyes. By now the water level has dropped below the bank nearest us and we can't see what's happening. We rush to get ready but quick as we are, nature is quicker. By the time we turn back around, the lake has disappeared: Where there was once an expanse of water a few miles wide and 32 feet (10 m) deep, only a few puddles remain. Ian, Alison, and I set off, more cautiously than ever, to investigate what has just taken place before our eyes. It takes us around two hours. We're all in high spirits and very excited: Our objective—the ultimate prize of our treasure hunt—is the sensor on the bottom of the lake (or rather, what used to be the bottom of the lake.)

We let our GPS lead us, a bit like metal detectorists sifting the beach, looking for valuable objects. Since the landscape has changed so much, there's an array of new brooks and streams across the surface, and we need to find a new way to reach our sensors. We also have to be careful: There could be new crevasses, meaning new dangers lying in wait. We're nearly there, all eyes scanning the surface of the ice before us, trying to get our bearings in a landscape that, just a few hours earlier, didn't exist.

Finally, we spot the *moulin*, the technical name for the hole in the ice through which the lake has vanished: It's a monster, incalculably immense. The ones I've seen or heard about before were no bigger than a few feet across. This one is at least 32 feet (10 m) from one side to the other. The sensor is lying a few feet away from the moulin. If I'd pressed the button a second later, it would've been swallowed up in the water flooding down the deep shaft to become nothing more than a tangible reminder of our presence—and a scientific relic embedded in the ice (and of very little use for any future data we might have been hoping to collect). The data we have collected, though, tells us that seconds after the lake imploded, the ice lifted by about 8 inches (20 cm) due to the water pressure at the bottom, which separated the ice from the rock and caused it to hydroplane and surge toward the sea. A block of ice half a mile (1 km) thick is nothing when up against the power of water.

Very cautiously, we make our way over. Alison hangs back; so does Ian. I'm the closest to the edge of the precipice created by the moulin. We're all attached via a safety rope and multiple harnesses, but the adrenaline and fear are intense. I peer over and see water cascading diagonally down the huge hole. My head spins just watching it, but I feel compelled to get closer to the edge. There, I've done it: I can see the walls of ice now—they're blue, solid, pure, and reach all the way to the ice bed. I'm lying on my stomach now, to get my head over the edge and see the bottom of what has become, in my head, a kind of fantastic creature, the entrance to the Cave of the Sibyl, or an underground tunnel as Dante might have imagined it.

The "hole" is around 32 to 50 feet (10 to 15 m) deep. For ice scholars like me, it's a truly memorable experience. The excitement about finally being face to face with an event you've only ever read about in books or seen in minuscule proportions is something that all researchers, whatever the field, share. To my sheer amazement, I realize the water gushing on the surface is not the main source of water pouring down this natural well in the ice: A horizontal channel at a depth of around 16 feet (5 m) is throwing out a high-pressure jet of water from one of the walls of the moulin and sending it smashing against the opposite wall. The technical term for it is an *englacial channel*, literally "channel inside the ice," something I'd read about but had never seen in real life, for obvious reasons. They form in the summer months and are initially fairly limited in size and number, although, as the season progresses and the volume of meltwater increases, they grow wider and wider until they join into a highly efficient drainage system that empties into the sea and raises sea levels.

In addition to rivers and lakes on the ice surface, water also flows in underground tunnels that change direction and size constantly. I've always felt this kind of subglacial labyrinth, under the blue mantle of the Arctic, has something in common with the Naples Underground—the network of catacombs that once, centuries ago, hosted a shadow city where life progressed in parallel, silently, to that of the people who lived above ground. The last time I walked through the narrow passageways beneath Piazza del Plebiscito, Via Roma, and the Spanish Quarters of Naples, I couldn't help but think about the canyons deep below the ice, like the one I'm looking at now.

The incredible thing is that, underneath these channels, there are even more of them—some even bigger and deeper. Data collected by NASA pointed to a previously undiscovered canyon about one mile (2 km) below the Greenland ice sheet. In some places, it's up to 2,630 feet (800 m) deep and 6.2 miles (10 km) wide, making it the longest canyon discovered on Earth to date—and bigger than the Grand Canyon.

In the end, we recover our instruments and head back to the others, which takes us several hours. The minute we reach them, we download our data and have a look. It's all there: From the sensor being released from the boat to the measurements of the depth of the lake, the fluctuations in temperature, wind, and whatever else changed the height, however imperceptibly, of the water column above our sensor.

More importantly still, we have data on the lake's collapse. It drained in around forty minutes. If we'd been there, we would've been swallowed up with no hope whatsoever of escape, making headline news as both a symbol of pioneering Arctic exploration and exceptional human stupidity. We begin our walk back to base camp, exhausted but ecstatic, hopeful that the data we've just collected will give us more insight into the behavior of this mysterious, majestic beast before us: the Greenland ice sheet.

CHAPTER 7

A Hole in the Ice

EUPHORIA BUBBLES BETWEEN US the whole way back as we excitedly bounce detail upon detail off each other, as if desperate not to let the incredible event we'd just witnessed slip away from us. We walk side by side, taking turns to lead the group, each time walking with someone new. When I find myself next to Christine, a veteran member of the project, our enthusiasm turns to reminiscing, and we end up laughing over memories of our first expedition together, in the Antarctic, quite a few years earlier.

Dry valleys—literally a row of arid valleys—is one of the few places in the infinite Antarctic expanse not covered in ice. Or not completely ice-laden, to be more precise. We were there with Christine to study a strain of bacteria that can only be found in that part of the world, and it was a truly unique experience. Access to the valleys is protected by an international treaty whereby no single country currently controls, or could ever control, the extreme southern continent. Only a few individuals are granted entry each year and only for scientific research. The landscape is lunar, Martian even, with red rocky peaks poking through ice still partially covered in snow. A violent and insistent wind like no other blows almost every day, eroding the rock and blowing rust-colored detritus over the surrounding land. Where ice meets sea, there's a brief glimpse of the emerald green of the icy ocean before the white of the sea ice takes over.

I remember we were walking back to base camp when we came across the carcass of what was probably once a seal. Christine explained to me that it was impossible to know how long it had been there, as the low temperatures and dry climate meant decomposition takes much longer. It could've been there for tens—hundreds, even—of years. Nevertheless, my most vivid memory of that trip is of the winds. Katabatics, they're called, formed by air rushing down from higher elevations, like avalanches. The force of gravity gives them a power that can often reach hurricane speeds. During that expedition with Christine, I was in my tent taking a well-deserved nap when I was awakened by what felt, in my sleepy state, like my daughter's hand brushing my cheek. It was only when I woke up properly that I realized it was the side of the tent billowing in my face. We took refuge double-quick in the only "steady" building we had (the kitchen tent) and waited in there for around ten hours until the wind died down.

No other place in the world has such jaw-dropping landscapes. But such uniqueness means that if you're privileged and lucky enough to come here to see them, you must also abide by the peremptory orders of the US government, which is "in charge" of that part of Antarctica. I remember the great pains we took to follow all the instructions and keep any impact of our presence on the continent to a minimum. Even just walking required the utmost care. We used the rocks and boulders around us like planks of an imaginary bridge over the expanse of algae, flowers, lichen, and other microorganisms in that ecosystem.

As we make our way across Greenland's ice sheets, I remain by Christine's side, breaking the unwritten rule of taking it in turns to be at the head of the caravan. I look around. We're surrounded by numerous tiny black holes, some only a few centimeters in diameter, others up to 4 to 8 inches (10 to 20 cm) wide. They're not unlike the ones Christine and I encountered in the Antarctic, and it's perhaps this memory that makes us stop to take a closer look. As we advance, we notice that more and more holes are magically appearing, and their edges are increasingly distinct. They're called cryoconite holes, and their morphology, distribution, and stability must be understood if we are to fully comprehend glacial ecosystems.

Dotted across the surface of the glacier ice, these cylindrical holes are an oasis of life, the only place where life grows on polar ice caps. Despite the waters surrounding Antarctica being home to abundant life-forms, there is very little life on the landmass itself—and bear in mind it covers an immense expanse, one and a half times the

size of the United States. It has the biggest freshwater
reserve on the entire planet (70 percent of the world's
fresh water) but is anything but hospitable. Temperatures
can drop to -195°F (-90°C), the lowest ever recorded on
Earth, and winds blow—"rocket" might be more accurate—at speeds of up to 155 mph (250 km/h), which is on
a par with the worst category-5 hurricanes ever to hit the
Americas and Asia. Greenland is not that much different:
All life on the island is confined to the few urban settle-
ments along the coast.

We observe these glacial, geometric structures together.
Peering through the water filling them, we see something
dark on the bottom. It's cryoconite, a sludge made of dust,
detritus, algae, and bacteria, found not only in the Arctic
and Antarctic but also in Canada, Tibet, and the Himalayas. It collects in these holes because the detritus absorbs
solar radiation, warming up the ice around it and causing
it to melt. One of the first and most fascinating things to
note about this detritus is that it's not just from planet
Earth: Studies have shown that every 2 pounds (1 kg) of
cryoconite contains roughly 0.35 ounces (10 g) of sand of
earthly origin and around 800 "cosmic spherules" (originating from comets, asteroids, or interstellar dust, and
part of the spectrum of materials that collide with Earth)
and 200 partially molten micrometeorites.

Even more spectacular is the fact that no one knew they
existed until a century and a half ago. Nils Adolf Erik Nor-
denskiöld was the first to describe them, the same man

who later set sail from Gothenburg on board the *Vega*, reached the Bering Strait, sailed around the northern coasts of Europe and Asia, and opened the famous Northeast Passage. That was back in 1870, and the Scandinavian geologist (he had dual Finnish and Swedish nationality) and Arctic explorer took it upon himself, and also had the honor, to be the first to publish a detailed description of the cylindrical melt-holes he witnessed in the ice (and which very few people, other than Nils and explorers like him, had ever seen).

Peering into the hole with my nose just centimeters from it, I ponder the doggedness of nature and its ability to surprise us, from penguins that cross the Antarctic peninsula solely to lay their eggs (when they would normally never leave their food supplies on the coast) to microorganisms like the ones I'd come across, and bacteria with names that sound straight out of *The Lord of the Rings*—roundworms, or the phylum Nematoda—whose survival depends on their digging into the ice. The ability of some of the "inhabitants" of the melt-holes to adapt to this natural environment and evolve under such extreme conditions makes them somehow unique and the ideal candidates for a study of extraterrestrial life. In early 2016, in fact, a group of Japanese scientists managed to "resuscitate" two microscopic animals that had been in hibernation for more than thirty years in ice samples collected in the Antarctic. Yes, they'd been in hibernation since Ronald Reagan's day (November 6, 1983, to be exact), when the US and USSR dominated the world and the very first skateboards were seen on the streets, and had "reawakened" from their long sleep on May 7, 2014, into a world of smartphones and social networks. The animals in question were *Acutuncus antarcticus*, a species of tardigrade—a microscopic (about 0.5 mm long) creature with eight legs, four to eight claws on each leg, and an odd appearance (like a tiny mammal with its fur removed).

Tardigrades have also become a veritable internet sensation of late, nicknamed water bears or moss piglets. Why are they so popular? Because tardigrades are a bit like video game heroes—you can freeze them, boil them, crush them, starve them, and they just keep coming back to life. There's no way to kill them! There could be no better candidate for a cryoconite hole. Water bears are one of the most fascinating creatures in the world for the way they can adapt to ultimate extreme environments. You can find them, for example, in the deepest ocean trenches or hottest deserts, on the frozen peaks of the Himalayas and—as we've seen—in the Antarctic, where Christine and I first encountered them a few years ago. They have succeeded in outliving dinosaurs and are so hardy that they can survive in extraterrestrial environments in addition to boiling and freezing ones.

The "defrosted" animals in the Japanese experiment had managed to survive through cryptobiosis, a process that causes their metabolism to slow to 0.01 percent of normal function. (Imagine your heart rate going from sixty beats per minute to only one every two minutes!) During cryptobiosis, all water in the body is released and the animal rolls itself up into a tiny, indestructible ball (the water in their bodies is sometimes replaced with a sort of natural antifreeze). Getting rid of water is the main priority, as it saves them, for example, from any cell damage caused by freezing.

The Japanese scientists described another, even more surprising tardigrade characteristic. One of the two animals (after being "defrosted") managed to reproduce. Here's what happened: The two animals—which the research team called Sleeping Beauty 1 and Sleeping Beauty 2 (SB-1 and SB-2)—were placed in two separate wells on a petri dish to be monitored and fed. An egg was then found as the experiment progressed. The researchers placed it in another well and called it SB-3. To each well they added agar (a jellylike substance used in molecular biology), bottled water, and chlorella algae (which contains chlorophyll). The ingredients were replaced weekly. More eggs were found over the course of time, all of which hatched safely.

There can be no doubt of the success of the Japanese team's research into the tardigrade, yet, as Christine reminded me, in this same context the work being done by a group of Italian scientists from the University of Milan-Bicocca is equally important: Working with colleagues from the University of Milan and the Bavarian Academy of Sciences and Humanities, they are applying DNA sequencing techniques for the first time ever to supraglacial sediments and to the "inhabitants" of the glacier surface. Their results show that the bacteria populating the glaciers of the Lombard Alps and Kashmir and the Baltoro Glacier in Pakistan use not two but four metabolic systems to produce what they need to grow. First, there's oxygenic photosynthesis (using carbon dioxide and solar energy to produce oxygen) and respiration (which uses oxygen and organic matter to produce carbon dioxide), then oxidation of carbon monoxide, and a photosynthetic metabolism that does not produce oxygen. Further studies have shown how the tardigrade genome contains more extraneous DNA than any other animal species known to man. To put it simply, instead of inheriting its genes from its ancestors, part of the tardigrade's genetic makeup may come from plants, bacteria, and fungi!

I direct my gaze to the bottom of the holes again, mesmerized, despite there being little to see with the naked eye. Tardigrades are not alone in the cryoconite holes. They share the habitat with other equally fascinating organisms, which is a major difference between Antarctica and Greenland. In the former, the icy dens can resist for years, coming through the seasons intact and becoming a sort of mini testing ground for extreme life. The ice cover stops the sun's rays from reaching the bottom of the cylindrical holes, with the result that photosynthesis cannot take place.

The existence of a different kind of life from what we are familiar with depends on a process known as bacterial chemosynthesis, as used by these life-forms.

Unlike photosynthesis, it exploits the energy generated in chemical reactions to produce organic substances. To put it more simply: These creatures are completely autonomous and self-sufficient, living their peaceful existence in complete isolation. Based on recent studies, the environmental factors in these landscapes and in these Arctic and Antarctic territories could be considered the closest to what we believe life would be like on other planets. Glaciers, especially the polar ice caps, are among the most extreme environments on our planet, not just because of the cold and isolation, but also because of the high levels of ultraviolet radiation, making them similar to planets or icy moons.

One of these would be Mars, or other icy celestial bodies like Europa, a satellite of Jupiter. Europa (discovered by Galileo Galilei in 1610) is a little smaller than the moon and mostly made up of silicate rock with a water-ice crust. The biological microsystems found in the ice are like "natural laboratories" helping us to understand alien life-forms. This is what makes them so important to astrobiology—the branch of science that studies life (or the possibility of life) on other planets in our solar system: NASA's recent mission to Mars, for example, or the renewed space race to explore our solar system in more depth. Or the many initiatives, some also in Europe, and the many stimuli (particularly money) from countries like China, which have recently been providing new and varied opportunities to dig much, much deeper.

I stop for a second to contemplate the sky. When we look up at the stars at night, enchanted by the unknown universe above us, we may often wonder if there really is life up (or down) there—a form of life similar to ours. At least we know that the next time we stargaze with a scientific eye, we can draw on what we've learned from these monster-like creatures, the true magic of which is undoubtedly the greater secrets they are yet to reveal—secrets that bind us all together on the planet we share, floating in the dark cosmos along with billions of other creatures.

This is not my only thought, however. Cyroconite holes accelerate the rate at which the ice melts because of the dark color of the admixture inside them and the increased solar energy they absorb as a result. The curious holes have a limited life span, given the role water plays. For example, the more the ice melts, the more life inside them can proliferate—but when the glacier in which the holes have formed begins to melt, the proliferating life inside them will be swept away by the meltwater current, and the holes will disappear. This is where climate change also plays a part: The more the glaciers melt, the more difficult it will be for the cryoconite holes to survive. And the life-forms discovered within them will become ever rarer; however, alternatively, it may be that the more meltwater there is, the more the holes will proliferate. The answer to this enigma is yet another mystery to be solved.

CHAPTER 8

Polar Camel

I T TAKES A WHILE, but we finally make it back to base camp. It's evening. It's been a long day, exhausting at times, and I don't realize how tired I am until we finally stop. It's satisfying to know we collected so much material, although it's only when you stop to look at the data calmly, safely inside, that you realize how useful all the instruments and equipment were that we worked so hard to bring with us.

On the way back, we traversed a lot of "old" snow, what people in our field call firn, the coarse, granular surface that accumulates on the top layer of a glacier. "Old" snow, to be clear, is the intermediate stage between snow and ice, the substance that gradually turns into the mesmerizing blue surface we walked across. We stopped regularly along the way to dig, as if looking for treasure. But when you think about it, the snow is our treasure, concealing within it Greenland's precious history.

The backpacks are heavy, but we can finally take them off, stepping out of the hefty yet necessary harnesses.

With every layer I remove, the comfort is so great I feel like I'm floating on the ice. Someone mentions dinner, but it's too early: We still have to sort out the equipment, check the instruments carefully for any damage, and store them carefully in the special cases that protect from the elements they're exposed to during the night, sitting outside our tents like faithful watchdogs.

I bend my knees to try to relieve the painful niggle radiating from my joints; standing up straight again takes more effort than usual, but it's worth it. I feel better already. When I turn around, Ian is standing on his head on the ice, legs in the air, in a yoga pose. Everyone finds their own way to get the feeling back in their bodies. I take a photo of him, chuckling to myself about how I could slip it in secretly to the closing slides of a presentation inviting the public to "see the world from a different angle." Unaware, Ian smiles at me with that gentle English manner of his that never leaves him, even when he's upside down.

Our hands hurt. They're covered in microscopic cuts even though we were wearing protection. The grains of ice in Greenland are not like the fluffy flakes we're used to on our mountains at home: They're different—bigger, tougher, sharper, more jagged. If you're not careful and leave a glove off or get a rip in one and touch the snow with your bare hands, you get punished for it. The cuts are minuscule— you don't feel them at the time because the cold instantly anesthetizes the skin, but the minute your hands warm up again, the burning feeling stays with you for days.

I go over and over the data we gathered. It's an old habit, a ritual, you could say. We scientists are always looking for something that doesn't fit—an attribute, a detail, something that asks a question or raises the tiniest of doubts. Ironically, we have more faith in doubt than truth. Scientists are perfectionists, obsessively so, and feel compelled to put every idea and every result to the test until we run out of breaking points (and when we can find no more, we finally accept, and feel grateful, that we have done something right).

Climate models are a good example of this. They're projections and a means of producing a rough estimate of the state of our planet—or better still, the processes that underpin it. Through them, we can try to understand what has gone before and what will occur in the future. They're a bit like a modern-day crystal ball, only digital and scientific. Climate models are one way we have of estimating, to a certain degree of accuracy, how much the temperature of Earth will rise if we continue to pump greenhouse gases into the atmosphere. We can predict what will happen to the glaciers, the forests, the snow, the clouds, and so on. With these models, we can explain and simulate what our planet was like in the now distant past, what it's like in the here and—sometimes chaotic—now, and what it will be like in the still hypothetical future. They give us irrefutable proof of what Earth was like millions of years ago when the city of New York as we know it stood on a thick polar ice cap.

There are those—perhaps among the climate change deniers—who claim that since climate models are incapable of forecasting with any accuracy the effects of a storm expected to hit our coastlines in a few days' time, how could they possibly simulate the behavior of Earth in one hundred or one thousand years' time? The explanation is easy and, once again, can be clarified with the help of an example: Imagine the climate model is the tool you use to predict when the water in a pot is going to boil. Thermodynamics (the science that studies how heat spreads and transfers) offers us an equation to make this estimate. No one would doubt that a scientist (or even a second-year physics or engineering student) could calculate this to a certain degree of accuracy. Yet, if we wanted to predict how each of the bubbles in the pot—one by one, with great precision—will behave when the water boils, we wouldn't be able to. Climate models are based on the same thermodynamic equations: They try to predict what will happen to the "system" Earth, not to individual bubbles.

Clearly scientists do not believe blindly in the perfection of models, in their data, or in anything else that has to do with science. For every model that works, there is another that fails. The worst thing we can do is to consider a hypothesis a foregone conclusion; we all do it sometimes, often in everyday life, when we accept a notion or piece of information as true merely because it's an accepted truth in our culture. In science, when observing or analyzing something, every time we say, "That's the way it is," or "It's a well-known fact," or trivialize and shut down an argument with an "Everyone knows that," we are making a mistake.

Solomon Asch's conformity experiments in the 1950s come to mind. Asch showed a group of people a cluster of lines and asked them to pick which one was most like a target line he showed them separately. Some of the participants were unaware they were taking part in a social experiment, believing it to be nothing more than a vision test, while the others were actors working with Asch. He changed the latter group's replies (from correct to incorrect) in order to influence the others and make them change their minds. By the end of the trial, many of the participants who had started out with the correct response switched to an incorrect one, simply to fit in with the rest of the group. This kind of behavior shows the degree of social compliance we're all susceptible to—a phenomenon that probably emerged when our ancient ancestors began to live in groups and communities, sharing their knowledge and expertise in order to find food and stay safe. Knowledge that was completely empirical.

Another story that clearly illustrates how collective beliefs can be difficult to debunk is the one about the polar camel. I heard about it on a talk show a while ago and have been fascinated with it ever since. We're so used to associating camels with the desert that the idea of a *polar* camel seems like the ultimate oxymoron. But reality can often be weirder than fiction, casting doubt on things we take as a given.

Let's go back to 2006. Natalia Rybczynksi, a Canadian paleobiologist and one of the most famous scholars and collectors of prehistoric life, is working as a scientific researcher at the Canadian Museum of Nature. One day, during one of her many expeditions north of the Arctic Circle, to an area rich in prehistoric fossils in the extreme Canadian tundra, she finds something on the ground. At first glance, it doesn't seem that interesting: It's rust-colored, fits into the palm of her hand, and looks a bit like a large wooden splinter. Natalia picks it up nonetheless, aware that the site is well known to scientists for its wealth of prehistoric plants. On returning to base camp, she thinks about this thing she's found. There's something not quite right about it, so she goes into the lab to examine it from every angle. Finally, she sees what's wrong: The mysterious object looks more like a bone than a plant. Over the course of the next four years, Natalia returns several times to the same area, each time locating additional fragments similar to the first one. She adds them to the collection until she has a total of thirty fragments of various sizes. A pattern begins to emerge, and she senses it could be extremely important. The problem she then faces is another: How to put all the pieces together?

There's no easy solution to the puzzle. It requires not only technical skill to piece together so many similar-looking and very tiny slivers, but also extreme manual dexterity, given how small and delicate the pieces are. Natalia trusts in the help of a three-dimensional scanner and manages to reassemble them digitally into a single piece. She's right. It's not a plant, it's a bone. More specifically, a tibia—and not any old tibia. To an expert eye like Natalia's, it becomes immediately obvious it's from a hoofed animal. It could be a cow, maybe a sheep. But a few of the particulars are telling her something else—about size, for starters: The fragment is too big to be from either of these animals. She runs some more tests and decides to try something new, something extreme: She slices into the bone.

And here's why. Here's the detail that sowed the seed of doubt: When Natalia cut into the tiny fragment of tibia, she instantly recognized a smell that was familiar to her, a smell she'd come to recognize after hours of anatomy lessons spent dissecting skulls of every kind. It was collagen—the main component of connective tissue, one of the four main types of tissue in animal bodies.

Those who work in the field will know—they'll probably have encountered it in the field as well—that connective tissue decomposes fairly rapidly as compared to geological aging (we're talking years or decades). In the case of the tibia Natalia reconstructed, something very singular had happened. The Arctic had acted like a huge natural freezer, and the bone fragments had been preserved perfectly, for how long no one knew, but it was as if they'd been placed in a powerful refrigerator.

Years went by and Natalia garnered more acclaim in the scientific world for the 2007 discovery she made with her team of a pinniped from the early Miocene period. A bit like a large prehistoric seal, the pinniped skeleton, or two-thirds of it, was rebuilt from bones found in Nunavut, the most northerly territory of Canada and forming most of the Canadian Arctic Archipelago. They jokingly named it "Bacon," after the much-loved rashers eaten widely in the Anglo-sphere and a staple in the Canadian researchers' diet during their glacial expeditions. The scientific name is obviously more serious—*Puijila darwini*—meaning "young marine animal" in Inuktitut, an Inuit language, with the second word simply an homage to Darwin.

The prehistoric pinniped opened up new opportunities (and new funding) for Natalia's studies and research, and it was also thanks to Bacon that the scientific world took even more notice of her. Many years later, the Canadian researcher bumped into a colleague at a conference. They chatted, debated, and shared data, and then the colleague happened to mention a new process called collagen finger-printing. This proved to be the turning point in the story.

Scientists had discovered that animal species each have different collagen structures, albeit minimally and often imperceptibly, giving them a unique fingerprint of sorts. Advances in technology had made it possible to compare the collagen structure of an unknown bone with those of known animals to identify the animal it belonged to. Natalia agreed with the colleague she'd met at the conference and the researchers in her own team to run collagen tests on the tibia she'd found in 2006.

The results were incredible. Not only did the bone date to a period of ancient history (three and a half million years ago), the prehistoric tibia was matched with an animal that no one would ever have said once inhabited a kingdom now covered in ice: It was the bone of a camel. Not the kind of camel we know today, clearly, because this one lived on the ice so long ago, and would have weighed around one ton and been almost 10 feet (3 m) tall. It was a distant relative, nevertheless, of the desert caravan variety we are more familiar with. Natalia's persistence had been rewarded with the discovery of the "Arctic camel," or, if you'd rather, the previously unknown "polar camel."

When we use the word *camel*, the first thing that comes to mind are the quadrupeds you'd have to travel to the Middle East or Central Asia to meet, including the dromedary (which, as you might remember from your school days, has only one hump compared with the Bactrian and wild Bactrian camels, which have two). Others, especially those familiar with South American camelids, may also be reminded of the Andean llama or alpaca. But no one would ever associate an animal like the camel with polar ice.

How did a camel end up at the North Pole? Scientists have known for years—and research over the past few decades has resoundingly confirmed it—that camels originated on the North American continent, where they lived for a large part of their many millions of years on Earth. Camels (and other more popular animals like horses, for example) migrated from the Americas to Europe and Asia over the Bering Strait, which, in prehistoric times, was a strip of land called Beringia—and probably the route humans took to reach the Americas.

The first camels (genus *Protylopus*) lived in North America around forty to fifty million years ago. According to some studies, during the late Oligocene epoch (from thirty-four to twenty-three million years ago) and the early Miocene epoch (around twenty million years ago), camels underwent rapid evolutionary changes that resulted in different species with different anatomical structures. They developed shorter legs and gazelle- or giraffe-like bodies. A richly diverse camelid family was reduced to only a few remaining genera in North America before they became completely extinct around eleven thousand years ago. The ancestors of the camels we know today descended from the now extinct *Paracamelus* or *Procamelus*, which crossed the Bering Strait into Asia around seven million years ago. Some of these camelids continued to live in the Americas, while the ancestors of modern camels adapted to their new lands.

To find the remains of a camel in the vast continent of North America would not be unusual for a scientist, paleontologists in particular, but no one had ever found (or ever imagined they might find) the remains of a camel so close to the North Pole. Three million years ago, during the geological Pliocene age, Earth was 2 or 3 Celsius degrees (about 4 to 5 Fahrenheit degrees) warmer than today— and sea levels were 82 feet (25 m) higher. In that period, the continents were different from the way they are now. North America and South America had joined together, and the resulting alteration in the ocean current caused the temperature of the Atlantic to drop; the subsequent moving together of Europe and Africa led to the formation of the Mediterranean. Despite this, as far as we're aware, the Arctic was still cold enough to have snowstorms, sub-zero temperatures, frozen lakes and rivers, and long spells in the winter when almost every day was in complete darkness. So our question remains: How could an animal like the camel survive in such an environment?

Natalia and her colleagues thought they had an answer. By studying the structure of camels in the modern world and drawing comparisons with what they knew of the polar camel, they came to a fascinating conclusion: The characteristics that enable camels to live in places like the Sahara Desert may be the same ones that were original-ly useful—in evolutionary terms—to overcome the Arc-tic winter. The wide feet might not have been for walking over sand and dunes, but to traverse snowy ground, a bit like primordial snowshoes. The famous humps, which, contrary to popular belief, do not contain water (they're actually filled with fat), may have been used millions of years ago to help the camel get through the long winter when food and light were scarce.

These are merely theories at the moment, although recent studies confirm they may be true—namely that several millennia later and after multiple migrations to warmer climes, camels did, in fact, begin to adapt (with wider feet, consolidating humps) to the desert habitat. It's a crazy thought, yet these attributes may be proof, or at most a strong sign, of their previous life in the Arctic.

This story not only reveals the migrant nature of the camel—or, more generally, the driving force of evolution—it also changes the way we think about the Arctic and the role it plays in the history of the planet. No longer is it just a cold and distant spectator. This discovery has placed it front and center in the theater of life. Above all, Natalia's discovery teaches us scientists (and not just scientists) the importance of thinking critically, whatever the situation, especially when dealing with ideas and beliefs that are assumed to be true.

Be warned, though: This kind of critical thinking has nothing in common with the mental shortsightedness and "denier" stances of those who reject established research for new paradigms entrenched in prejudice. It is quite the opposite. Critical thinking encourages open-mindedness and tolerance of new and different ideas, and makes us all more ready and able to accept the possibility of "new truths." Whether we're discussing science or philosophy, religion, or history, we should remember Natalia and the story of the polar camel: how an apparently microscopic fragment, which could so easily have been overlooked, managed to overturn even the most deeply entrenched beliefs.

These were my thoughts as I gazed out at the immense ice sheets after saving all our data. Ignorance can be overcome. All it takes is to let go of "absolute truths" and leave room for research—always and everywhere.

The world is a place of constant change, where everything flows—especially ideas.

CHAPTER 9

The World Through a Lens

THE DINNER RITUAL HAS BEGUN. The first necessary step is melting ice to get water to cook our food. We light the camp stoves set out on an aluminum table—a simple piece of equipment, yet just as important as our sophisticated instrumentation—the legs of which we've wedged into the ice to keep it as level as possible. The water left in our drinking bottles is poured into the bottom of the pot where we'll be melting the ice. It's an old camping trick to make the ice melt quicker and make more efficient use of the gas reserves in our camp stoves. Liquid water transfers heat from the flame to the ice more efficiently, pushing the hydrogen and oxygen molecules around in their molecular embrace and making them jump faster, acquiring kinetic energy.

I stare at the piece of ice we extracted out of the ice sheet, and which is now sitting in our pot. I wonder when that particular piece of ice was formed. It's difficult to know exactly because ice moves, flows, mixes with other ice. It's a living entity. It may have formed when Hannibal crossed the Alps or when the first humans (first hominids, that is) began their long march across the continents and kicked off the colonization of Earth. I watch it slowly melt and for a second imagine that the memories it held are now melding with the water in a sort of bizarre time machine. The chemical compounds we normally use to date something fuse before my eyes in a thermodynamic dance, intoxicated by the energy they receive from the flame— losing their identity forever.

What about me? How many memories of other lives— some dreamed, some lived—are forged within me?

Patrick touches my shoulder, snapping me back to reality. He's smiling. He tells me the piece of ice in the saucepan reminds him of a telescope lens. He breaks off a smaller piece and looks at me through his imaginary telescope, his smiling face distorted behind the icy tablet. I laugh, too. Does he know, I ask, that polar ice was actually used in the past to observe the universe? He looks at me, surprised, interested, and a little confused. I nod and sit down: I might be tired, but since I've said this much, I can't back out of explaining now. After all, it's not like we don't have time before dinner.

Many of the telescopes that have given us the most recent, most accurate information regarding the birth of our universe were able to do so because they were located either above or below the polar ice caps. Yes, *below*. But first things first.

Greenland has the honor of hosting, at the Thule Air Base on its northwest coast, a radio telescope with a 40-foot (12 m) radio antenna. It was installed in 2017 and is part of the global Event Horizon Telescope (EHT) network, which also includes the famous Atacama observatory in Chile. EHT is an ambitious international project whose ultimate objective is to study black holes. The fact that we can't "see" a black hole is a huge problem for physicists because of the very nature of black holes—the immense gravitational attraction from which no form of electromagnetic radiation can escape. Unsurprisingly, to learn more about them, astronomers try to study their shadows.

The EHT project will generate images of two large black holes: one in the middle of our own galaxy, the Milky Way (yes, if you didn't know already—we have an enormous black hole in the middle of the Milky Way), and another, bigger black hole in the center of nearby galaxy M87. Other telescopes in Chile and Hawaii point in the same direction, and data will be pooled from all of the telescopes in the EHT project to produce the final images. Along with data from the Antarctic telescope, in 2019 humanity accomplished something previously thought impossible: A photo of a black hole was captured, along with all the consequences (scientific and other) of such a discovery.

Why Greenland? Patrick's question is a perfectly valid one. There's a reason Greenland was picked as a location. To capture the desired images, the telescope has to be in a place where the atmosphere is very dry. On a clammy, damp day, you can't see past the neighboring houses. But if it's windy and dry, you can easily see places a long way away. There aren't many places on the planet with the right type of climate, but high up on the summit of the Greenland ice sheet is one of them.

Events on the other side of the world are even more fascinating, more specifically at the South Pole Telescope (SPT), 33 feet (10 m) in diameter, located at the Amundsen-Scott South Pole Station in Antarctica. It is designed to observe the universe in the microwave (yes, the things that heat our food quickly when we're in a rush) region of the electromagnetic spectrum. The design goal for this telescope is not to study black holes but to listen for any remaining noise from the explosion that created the universe, that is, the big bang. The theory regarding the origins of the universe tells us, in fact, that during the big bang the universe was condensed into a tiny dot, smaller than an atom. The only thing we can say with any certainty is that—in the very early stages—it was made up of a concentration of incredibly hot energy in which time and space, as we know them today, did not exist. This stage lasted a time so short even our most sophisticated instruments can't measure it: 10^{-36}, meaning 0.000000000000 00000000000000000000000001 seconds. To give you an idea, the smallest unit of time measurable with a modern atomic watch (the most accurate timepiece in the world) is *only* 0.000000001 seconds.

Several stages followed this first one, starting with inflation—when our universe expanded exponentially—after which it cooled, before progressing finally to the gravitational stage when matter collapsed as a result of gravitational forces, resulting in the formation of stars and galaxies over the course of billions of years.

The inflation stage is where the Antarctic telescope comes in. According to a number of researchers, during the inflationary stage the cosmos was shaken by a myriad of gravitational waves. These waves caused the universe to vibrate as they passed through it, a bit like the skin of a drum when hit or a guitar string when plucked. Echoes of this passage could be compared to the shock waves of an explosion that continue to propagate, or like the waves that ripple out from a stone when you throw it in the water.

One way of testing the big bang theory is to prove the existence of gravitational waves, which means finding hard experimental evidence. Clearly this sounds like an impossible endeavor, but let's not forget that science, throughout its history, has often succeeded in tasks that seemed impossible at the time (flight, computing, space exploration, genetics). The key to the mystery seemed to have been finally found a few years ago. In 2014, a group of researchers announced that they had uncovered primordial gravity waves using two telescopes in the Antarctic: BICEP2 (Background Imaging of Cosmic Extragalactic Polarization) and the Keck Array. The news traveled quickly around the world, not least because the discovery had laid bare one of the most remote secrets of our universe but also because it confirmed many of the hypotheses of modern physics, the big bang theory included.

Alas, it was all an illusion—a galactic mirage destined to quickly fade. The announcement soon became a detraction, and, in a matter of months, the discovery that was supposed to be the biggest of all time was duly overturned. Only a few weeks after the "big discovery," a European team of researchers using data from the Planck satellite (in honor of Max Planck, German physicist and forefather of quantum physics) showed that the BICEP data was not an echo of the inflation of the universe but rather a signal generated largely (if not entirely) by dust in our galaxy that emits a signal similar to the one their colleagues had been looking for. What a pity the whole project subsequently came to an end.

I stop: A gust of wind swishes through the camp, breaking the hush that has fallen. I look intently at Patrick, who gazes straight back, waiting for me to resume. I smile. To me, the ice has always been more of a lens looking inward than a technological tool, because when living and working out here in this icy domain, the soul ultimately melds with this seemingly rigid, inert material, and the ice becomes the lens through which we contemplate our own thoughts, fears, and challenges—a lens revealing the many facets of our personality. In my case, this would be the desire to isolate myself, a trait common to many glaciologists who are known to be temporary deserters of a society that is too fast, too chaotic, and too noisy for them—as well as seekers of the kind of intimacy that only a place like this can give you. Then there's the eternal curiosity, the determination to come this far, to a place where no one has ever set foot before—and even enjoying the logistic, physical, and mental preparation it entails. Here, where silence reigns, you have space to think your most intimate thoughts, words waiting silently to materialize, as if written in invisible ink and suddenly taking shape when brought close to a flame. Afterward, when the battle's over, the ice bolsters my spirits—stretching my soul in directions I would never have thought possible.

Patrick wakes me up again from my reverie with a joke. I give the water in the pot a quick glance and go back to my cataloging of ice telescopes. Personally, the one I find the most interesting is the IceCube Neutrino Observatory, again in Antarctica, at the Amundsen-Scott South Pole Station. If you want to visit it, don't go looking for the observatory on the surface, because you'll never find it. Like a lot of its peers in the Far North and South, it is embedded deep in the ice, given that it is designed to go digging for neutrinos. Neutrinos are elusive particles with no mass and no electrical charge, meaning they barely, if ever, interact with matter. Unlike other charged particles that interact with the magnetic fields of stars and other celestial bodies, neutrinos pass through objects unaffected, traveling in a straight line from their source and making it difficult, if not impossible, to detect them. Despite there being such significant numbers of them (around fifteen million neutrinos pass through a square centimeter of our bodies every second), it is this elusiveness that led to the need for a complex neutrino detector.

To capture these particles and learn more about where they come from (which is just as important), physicists need gigantic detectors installed—so big they're measured in kilometers—in material that must be "optically transparent"—that is, ice. In this detector, more than five thousand optical modules were inserted between eighty-six shafts sunk 1.5 miles (2.5 km) deep into the ice. Only clear mediums like water or ice can be used to detect neutrinos. When a neutrino smashes into a proton or neutron in an atom, the resulting nuclear reaction produces secondary particles that emit a flash of blue light known as Cherenkov radiation.

The sensors in the ice can detect with great precision when a neutrino passes through and interacts with other terrestrial particles, which can complicate the readings. Originating in the atmosphere, these particles trigger the same or similar mechanisms as neutrinos and confuse the computer systems, whereas when the sensors are sunk deep into the ice, the ice itself eliminates any interference that could contaminate signals coming from neutrinos, before they reach the detector.

Patrick moves away and I'm finally alone. I shut my eyes and think that right now, in this place, sitting by a pot of water full of ice and not much else, billions of neutrinos generated millions of light-years away—as old as the stars, the planets, the galaxies—are passing straight through me.

I imagine I'm one of them: a neutrino. Free. Weightless, massless, chargeless. I look around; there's nothing like sitting under an infinitely big sky and feeling infinitely small.

CHAPTER 10

Northwest Passage

DINNER'S READY. TIME, AT LAST, to relax, clear our heads, laugh a little, and think about the events of the day. But the laid-back vibe around the table doesn't last long: A beep from the satellite phone interrupts the pleasantries with some unpleasant news. It's a short but no less disquieting message. The fabled Northwest Passage, the route from the Atlantic to the Pacific Ocean along the northern coast of North America, has reopened. The sea ice that once prevented ships from transiting has melted—a victim of global warming.

Celebrated in novels and films, this sea route through the Canadian Arctic Archipelago has been at the forefront of a saga alternating between ambition and tragedy going back to its origins in 1523, when King Francis I of France asked the famous Italian voyager Giovanni da Verrazzano to explore the coasts of North America—from Florida to Terranova—while aiming to find a sea route to Asia. Further voyages were made in the years that followed: Jacques Cartier explored Terranova and managed to reach the Saint Lawrence River; John Davis explored the Cumberland Sound off Baffin Island; in 1609, Henry Hudson sailed along the river that now bears his name.

Only two centuries later, the legendary Malaspina Expedition set off from Cádiz with two ships—the *Descubierta* and *Atrevida*—and traveled from one end of the world to the other, crossing the Atlantic in a mere fifty-two days, down the coast of North Africa, around Cape Horn, up the coast of South America, and northward to Panama until finally reaching Alaska. However, the expedition had begun in 1789, and France was in the grip of a revolution that was to change the fate of the world—and unfortunately also the fate of Alessandro Malaspina. Shortly after returning to Spain in 1794, he supported a failed coup and saw his voyages of discovery end in a Spanish prison.

The great Age of Exploration had almost come to an end when Sir John Franklin was instructed to sail two warships, HMS *Erebus* and HMS *Terror*, in search of undiscovered lands and the much fabled "passage" through the Arctic (in 2018 the subject of an AMC series called *The Terror*). The captain hoped to avoid the Arctic ice jam by transiting Baffin Bay but may have underestimated the danger of such a passage: The ships ran aground near King William Island, trapped in the ice. The entire expedition died. Nearly two hundred years later, in September 2014, sonar images finally located both wrecks: *Erebus* first, then *Terror*.

We look around at each other. For days we've been on edge, waiting to see how the centuries-old battle between the ice and climate change in the Arctic will play out this year, unsure what will become of the once thick, impenetrable barrier that has been gradually retreating, leaving room for the immense, ice-cold ocean to swirl through. The immediate consequence of this will be an influx of vessels keen to brave the dangers of a sea that hasn't been navigated for hundreds of years, and take the shortcut from the coast of Alaska on one side to the East Coast of the United States on the other, instead of taking the usual route through the great feat of engineering that is the Panama Canal. The Arctic—the promised land of modern shipping—is this century's Panama Canal.

There might be space to squeeze through the ice, but the journey is still fraught with dangers, both known and unknown: Not a lot is known about the seabed along the various sea routes, especially areas of the Canadian archipelago, which have only recently come to light (often only partially) as a result of the changing sea ice on the surface. Sea ice arises as salty seawater freezes, and it is the most common form of ice in the Arctic. During the winter months, there would normally be a permanent cover of ice across most of the sea, whereas in the summer months, this cover cracks and melts, creating an archipelago of ice chunks that break off the pack ice and are driven together into a single mass as they drift on the sea. Icebergs, even small ones, can cause devastating damage to ships and people. Anyone attempting to navigate these waters should keep the much-fabled, heartbreaking story of the *Titanic* in mind as a warning. Modern ships have much more sophisticated and modern equipment to keep them safe, but would they be enough to navigate with your eyes shut? Not a chance. Even the satellites keeping an eye on the situation from above can't detect tiny icebergs.

The text message has livened us up and made us chattier than usual. Someone mentions the unknown dangers of the Arctic for those who dare to brave these seas. Someone else points to the hidden risks we ourselves are subject to, traveling overland. In effect, one of the biggest hazards for explorers in this part of the world is the lack of any kind of marine support in the event of an emergency. This would make search and rescue an expensive business, possibly even impossible. Anything could happen, from a bad toothache to the flu, a sudden illness, a fall, a broken leg, or a frozen foot. An accident or unexpected disaster can come in many forms, but the hope of a successful rescue operation still remains extremely remote.

There are multiple aspects to consider: the dangers to the ship, to the crews, and also to the environment. Oil spills, even the tiniest, can do immeasurable damage, given the very sensitive balance between the Arctic's ecosystems, biotics, and sea. At this latitude, a mechanical fault of apparently little importance could trigger an environmental disaster of monumental proportions and end up altering—probably permanently—the Arctic as we know it today. I'm not just talking about oil spills; even ship noise has the potential to alter the ecosystem of the Arctic Ocean, interfering with the communication systems of whales and other marine mammals inhabiting these waters and frequenting the same waters that cruise ships pass through. Ian and Christine remind us that the opening of the Northwest Passage is, in itself, a sign of how climate change is clearly affecting the Arctic, while also setting in motion events that could be lethal for the ecosystem. Navigating through the Arctic can shave thousands of miles off journeys, but that savings comes at an unimaginable price.

Patrick is one of the quieter, more reserved members of the expedition, but this conversation seems to have roused something in him. He begins to tell us about an article he read on Arctic tourism recently. We're confused, but the fact that he's decided to share it with us gets our attention. The gist of his point is crystal clear, though, and very frightening: The number of people visiting the Arctic is rising steadily from year to year.

The numbers tell us that the Arctic is still a little-known area, which suggests what the arrival of tourism could mean for this region of around 2 million square miles (5 million square kilometers) inhabited by a total of thirteen million people (the population of Tokyo), divided among nine different nations: Canada, Denmark (with Greenland and the Faroe Islands), Finland, Iceland, Norway, Sweden, Russia, and the US. The indigenous population totals around nine hundred thousand from twenty different peoples, including the Inuit (the most well known), the Yakuts from the Siberian Arctic (the most populous), the Aleuts, the Yupik, and the Tungusic peoples.

To this world come the tens of thousands of visitors flocking to the region seeking the thrill of ice tourism. There are no specific figures yet to quantify the extent of the trend, partly because the Arctic is not your usual continent with plenty of data on every aspect of its nations, cities, trade, and climate. At any rate, one thing is certain: Most tourists arrive in the region on "polar cruises."

The desire to go on pleasure cruises to the Arctic is nothing new, given that the earliest voyages of this kind can be traced to the late nineteenth century, when US industrialist William Ziegler rented ships to take whalers to Spitsbergen in the Svalbard archipelago. By the early 1900s, Ziegler—who made his fortune with the Royal Baking Powder Company—had developed such a strong interest in polar exploration that he decided to devote two of his ships to it, the *America* and the *Belgica*, which could also be boarded (for a hefty ticket price) by passengers who were not part of the scientific crew. It was an unmitigated financial disaster, yet proof that such a thing was possible.

Nowadays the route is more easily navigable, and Norwegian and Russian cruise ships are doing the lion's share of the work. More recently, however, the Canadians have been making inroads to secure themselves a share of what most observers see as the biggest tourism market of the fuure. Intrepid tourists are now offered the option of an all-out, two-week "polar adventure." Adventure with creature comforts, of course, and scheduled connections. A charter flight brings them to the Svalbard islands (via Oslo) from where they are transferred (by plane again) to Camp Barneo, a base located 89° north, roughly speaking, because the camp disappears every year when the ice melts and has to be rebuilt the following spring. The camp hosts an airstrip, general facilities, and a handful of helicopters. Arriving there, tourists have two options: to travel to the North Pole on skis (a week's trip) and sleep in tents, or, for the lazier, make the journey by helicopter—out and back in a day, with dinner in a tent included, as well as a satellite phone call to share the news with family and friends back home. Trips like these, even those that go halfway to complying with environmental restrictions, can very quickly become a huge problem when visitor numbers rise.

I fall silent, thoughts crowding my head. Patrick keeps talking, sharing information. Summer 2016 was a milestone for Arctic navigation: In August of that year, *Crystal Serenity*—registered in the Bahamas and owned by Crystal Cruises—sailed from the port of Seward in Alaska to New York with one thousand passengers on board. It sailed along the coasts of Canada and Greenland, completing the 930-mile (1,500 km) voyage in thirty-two days. A dream that had been building for centuries had finally become a reality.

In previous attempts, the cost to human life had been too great. The earliest known forays north to open a new sea route were made beginning in 1497 by John Cabot, who was, according to records, born in Gaeta but an honorary citizen of the Republic of Venice, having come close to becoming the city's very own Christopher Columbus of the northern seas. Unluckily for the navigator, Venice's doge had no interest in undertaking exploratory sea expeditions, so Cabot (following a period serving the king of Spain) had his dream funded by English King Henry VII. In a 1497 voyage, Cabot "discovered" Canada. It is reported that, on his second expedition (1498), Cabot made landfall somewhere in the United States, possibly at the latitude of New York. But in a mix of fact and fiction, it's not fully clear if the Italian explorer made it or not (considering he thought he'd arrived in Asia when he landed on the Canadian coast). The first explorer proven to have completed the journey, a good four hundred years later, was Norwegian explorer Roald Amundsen. On a voyage that lasted three years, from 1903 to 1906, he successfully navigated the Northwest Passage, from Baffin Bay to the Bering Strait. There have been a further 240 traverses since then. Most of these voyages—17 of which were in 2015—were made after 2007, when, for the first time since the Arctic was monitored with modern scientific instrumentation, the passage was fully navigable in summer.

In the meantime, our soup is getting cold and the steam rising from it has dwindled. Sitting down together to a plate of piping-hot food is one of the more enjoyable parts of life on the ice, and the aroma of the herbs and spices adds flavor to our conversation. Soup is an Arctic delicacy. For obvious reasons of weight and space, we bring the bare necessities out with us. But this soup couldn't be more different from those of our early expeditions. Over the years, we've learned to make it more flavorsome and pack out the calorific value, using ourselves as guinea pigs: spices, olive oil (a staple for us, we always take it, no matter how much it weighs), butter, rice, pepper, cheese, and there you have it. As we scoop what's left onto our plates, Patrick resumes his story.

Crystal Serenity might've earned the record as first passenger vessel to sail through the Northwest Passage, but in truth the MS *Linblad Explorer* managed to navigate it as early as 1984. *Crystal Serenity* was by far the longest ship—coming in at 820 feet (250 m) in length and almost 69 tons (63 metric tons) in weight—ever to attempt (and successfully complete) this epic journey through the Arctic. As far as sea navigations go, this is not to be underestimated, because size can be important, in many cases crucial, when maneuvering through a labyrinth of waterways—90 percent of which are completely uncharted. As I mentioned earlier, the potential hazards are many: low-lying rocks, unpredictable currents, unmapped inlets, ice floes that resisted the thaw. To complete the passage, *Crystal Serenity* required an icebreaker escort vessel and two helicopters, which scanned the surrounding waters constantly to detect drifting ice.

By the way, how much would such an extreme cruise cost, given the intricate organization required to undertake it? Patrick gives a sad smile: The cheapest ticket was a mere $20,000, whereas passengers preferring the convenience of a luxury suite shelled out up to $100,000. Excursions were easily another $5,000 per person on top of that. The cruise sold out in just three weeks and—surprise, surprise—the passengers were among the mega-rich in society. There are, sadly, parallels with big-game hunting, where the elite of international tourism pay top dollar for the right to hunt species at risk of extinction: rhinoceroses (white and black), wild camels, Asian elephants, tigers, fin whales, hyper-protected pandas, and even polar bears. The Arctic is a different kind of business, but the damage is the same. Strict rules and restrictions have now been introduced in Africa and parts of Asia, but the legislation in this part of the world is still unwritten.

The increase in sea traffic through the ice can also be seen in the steady rise in accidents recorded. The numbers emerging from insurance company data are highly disturbing: In the Arctic Circle, around seventy accidents were recorded in 2015 alone, 30 percent more than the previous year. Alison interrupts with news of a cruise ship carrying around two hundred passengers—a lot smaller than *Crystal Serenity*—that ran aground roughly 55 nautical miles from the coast. It took more than two days for the Canadian Coast Guard to get everyone, both passengers and crew, into lifeboats and to safety. The wreckage was abandoned, left to be swallowed up by the ice.

On the subject of the human impact on polar regions, Ian draws our attention to an alarming discovery made recently in the Antarctic. Traces of caffeine, acetaminophen, and cocaine were found in the ice at the South Pole, at levels similar to those found in densely populated areas of Europe. The high concentration of such substances is most likely due to their overuse by the hordes of tourists, around a thousand every year, now regularly visiting the Antarctic. Wastewater and effluent from ships is so saturated with these substances that the oceans into which they are freely discharged have higher and higher concentrations—with devastating consequences for the environment.

Tourism is not the only problem, though, I remind my dining companions. There's also the "the Polar Silk Road" project, named in recognition of China—the non-Arctic state who's behind it. Launched in early 2018 in an Arctic policy white paper, the Beijing government plans to halve the time required to link China to northern Europe (twenty days instead of the current forty-eight) by developing new shipping lines via passages that have opened up by global warming, taking its place as a "polar great power." Some have their misgivings and are concerned that the prize they are actually after are the "hidden treasures" of the Arctic.

Under the region's land, water, and ice lies a veritable treasure trove of oil and gas reserves—a hundred billion tons of
them, to be precise, equal to nearly a quarter of Earth's natural fossil fuels—all, or nearly all, of which are still untapped.

The Arctic is home to 13 percent of Earth's unexplored
oil reserves and 30 percent of untapped natural gas. Then
there's gold, silver, iron, uranium, and diamonds. According
to the United Nations Convention on the Law of the Sea
(UNCLOS), all nations can make claims to the Arctic and
attempt to extend their sovereign rights to the outer edge of
the continental shelf stretching from their respective coastlines, extending their exclusive rights over all resources in
the seabed and sub-seafloor. If journey times are brought
down and the investment made more financially viable,
China could shift millions of people over the northern sea
routes, more than the Arctic has seen in its entire history.
Worse still, it could quickly turn into a new gold rush where,
instead of courageous miners traveling to the Yukon seeking
their fortune among the gold reserves, we'll see world super-
powers scrambling to get their hands on the polar ice—with
catastrophic, irreparable consequences for all of us.

Ian watches me attentively, raising his eyes, chin tucked
into his neck as if to trap in the heat his body is emanating
under the many layers that are protecting him from the
extreme temperature. The lines of fatigue on his face are
getting deeper—as they are on everyone's face—but it's a
happy sort of fatigue because it's accompanied and aided
by the enthusiasm we've felt since we first arrived—and

which drives us on, day after day. The conversation continues and our faces soften, relaxing. Christine is putting the final touches on our main dish—beans with meat and herbs. It's not bad, considering that, after a day on the ice, our stomachs would accept anything that comes their way, regardless of how good or bad it is.

The opening of the Northwest Passage, as I was saying, is not an issue purely for the influx of tourists it will bring. Greenland's economy depends mainly—apart from fishing, obviously—on the mining of rare earth elements and uranium. Rare earth elements are seventeen chemical elements near the bottom of the periodic table that are not valuable in small quantities (like gold and silver) but are widely used in military applications, renewable resources, and consumer technology, since they are crucial to the manufacture of smartphones, fiber optics, hard disks, hybrid vehicles, microwaves, and superconductors.

The first rare earth element was discovered in 1787 by Swedish chemist and army lieutenant Carl Axel Arrhenius, who extracted it from a mine in the village of Ytterby on one of the many islands in the Stockholm archipelago. He came across a black mineral he'd never seen before and, believing it to be a new element, named it ytterbite, after the location where he'd found it. More elements have been identified since then, each given equally obscure names: yttrium, cerium, lutetium, lanthanum, scandium. I happen to glance at the satellite phone, and it occurs to me that there's probably a piece of Greenland right there inside it.

Well, let's just say we've brought it home for a visit.

With the retreat of the ice in recent years, more and more bare rock has been left exposed, rock that was once covered, hidden for thousands of years. In turn, the shrinking ice sheets have revealed once-hidden deposits of rare earth resources. As often happens in cases like this, the benefits are not reaped by the local townsfolk, but by big multinationals and the mother country, Denmark, who have both means and method to extract and export the precious minerals. It's a genuine tragedy, especially when you think about how poor northern Greenland is: The sun almost completely disappears in winter and the people are clustered in tiny villages of a few hundred people, with soaring alcoholism and sexually transmitted diseases and no nearby hospitals. I feel the same anger for this injustice as I do when I read stories of how indigenous American civilizations were destroyed by European conquistadors arriving in South America or by settlers in North America. Ancient cultures are slowly being annihilated, not just physically but also economically, and by social stratification.

As our plates fill with beans, a strange silence weaves through plates and forks. Silent thoughts fill our heads: the hope that, one day, our work could somehow be of help to the local people at the same time as the realization that some of the data generated by our research ends up in the hands of foreign companies investigating how, where, and when to start drilling—and which glacier will reveal yet more of the bedrock from which to extract mineral resources, as it accelerates its slow slide toward the sea. We are not humanitarian workers; we are here as scientists to study the impact of climate change, yet reminding ourselves of this doesn't help to alleviate the sense of guilt.

Christine tries to cheer me up, reminding me of the new hydroelectric power plant that Denmark built in Ilulissat, in southwest Greenland, not far from here, which is driven by glacial meltwater. I nod, thinking back to when I witnessed Danish helicopters transporting building materials for the plant. It was like a scene out of the Pink Floyd film *The Wall*, the bit where steel pylons fly over enchanted landscapes that are still pure and unspoiled. My mouth puckers in a bitter grimace: None of the people I met or saw working there were locals.

Ian launches a new topic, which is no less worrying than the previous one: What would happen if Russia and the United States (whose economic and political interests, as we've known for years, are diametrically opposed) were to embark on a silent, invisible war, much like they did in the Cold War era, with the intent of damaging the progress each side is making on their respective Arctic coastlines? The question is anything but outlandish or offhand. Putin's Russia has an active interest in the Arctic, in part because of the numerous military operations—submarine maneuvers for instance—performed off the Siberian coast and in part on account of the oil reserves under the Arctic Ocean. A few years ago, the Kremlin launched its ambitious (and according to some, also unrealistic) Project

Iceberg to explore the harsh, icy sea and build platforms to extract the underwater oil. An enormous 600-foot-long (182 m) nuclear submarine called *Belgorod* will carry out underwater surveys and lay communication cables under the ice, as well as acting as a mothership for a flotilla of smaller submarines.

One new weapon that may be wielded in this "ice-cold war" is geoengineering, the ability to alter or create meteorological events (even transforming the climate) by releasing substances into the atmosphere to deliberately trigger or reduce the formation of clouds and, in turn, preserve or melt ice. Clouds can shield the ice from solar radiation—one of the main causes of melting. In itself, geoengineering is not necessarily a threat and can be used for both scientific gain and as a public service. The cause for concern is that there are no specific laws governing geoengineering activities (scientific or otherwise) that target sea ice. The only applicable ones at present are those regulating territorial claims to international waters, up to 62 miles (100 km) from the coast. This means that Russia and the United States—or any other nation with the capability—could affect the climate of another country, with obvious geopolitical implications. It's a terrifying prospect (and not just because we're in Greenland).

Dinner winds down gently, as does our chatter. It's time to go to bed. The nights on the glacier are long, although there are times when you're wrapped in the warmth of other humans that they feel far too short.

CHAPTER 11

Freedom

STANDING UP IS HARDER THAN USUAL. We're heavier after the meal, and the camping chairs we use make you sink so far down they're basically uncomfortable. It's just me and Patrick left; the others are heading back to their tents for some well-deserved rest. Ian and Alison giggle companionably as they brush their teeth and look out at the breathtaking view. It's not a cold night, not for these parts anyway, and the tent will be warm inside, as I mentioned earlier—sitting in the sun all day, the air inside heats up so much that if we were to untie the guylines keeping it anchored to the ground, it would lift up like a hot air balloon. You can't get complacent, though, as from one minute to the next the weather in Greenland can change, switching from cold to freezing in the blink of an eye. It's happened to me before, even during this expedition: You go to bed on a mild night and wake up a few hours later in the thick of a freezing fog, clothes sodden and bones aching as if impaled by glacial daggers.

We clear the last few things from the table. There can't be anything lying around, or a sudden gust of wind would simply sweep it all away. Everything must be in its place, sealed, airtight, watertight, organized, and packed away. I look around again; I'm tired but don't feel like sleeping yet, still wired from the day's events. I sip the scotch I brought for moments like this. It warms me and I think about how this war we're fighting is going to be a long one—the war on climate change, on humankind's devastating impact on the natural world and the uncontrollable, unchecked progress for which the environment is paying a hefty price. It's a war that will be long and very, very difficult. No one knows how it might end.

As we look to the past, we find more and more clues as to what the future will hold—and when we read them carefully, we have serious cause to be concerned. Deeply concerned. The answers we seek are all there in the data. Take sea levels, for instance. When temperature and carbon dioxide levels in the past were similar to those recorded now, sea levels on our planet were much higher. Some say only a few meters, others say closer to 32, even 50 feet (10 to 15 m). Hundreds of thousands of years ago, changes occurred much more slowly, at a "geological" pace, as part of a natural cycle. It took thousands and thousands of years to get to the carbon dioxide levels that we have now equaled and surpassed in just two and a half centuries (basically since the first industrial revolution). In the warmest period *Homo sapiens* have ever known on planet Earth—known as the Riss-Würm, or Eemian, interglacial period, between 130,000 and 115,000 years ago—the global climate was completely different. Hippos, monkeys, elephants, and lions roamed in England (as shown by bones found in London's Trafalgar Square in the 1950s), whereas now they are only to be found in the tropics. Sea levels in this period were between 20 and 30 feet (6 and 9 m) higher than they are today, and temperatures a full 2 Celsius degrees (about 3.5 Fahrenheit degrees) higher than they are now. I can assure you, this difference is a lot—and it brings extreme, if not deadly, consequences for fauna and flora; 2 Celsius degrees is also the limit scientists have set as their red line in the sand, beyond which it will be too late, they say, to save the planet. Global warming has currently reached one Celsius degree over preindustrial levels, and twice that in the Arctic. According to recent estimates, the sea is predicted to rise more drastically in the next few decades as the planet reacts to the increase in greenhouse gases.

It's difficult, nearly impossible, to say what will happen exactly, and even if the forecasts available to scientists are not 100 percent certain, many predict that the worst is yet to come. It is said that sea levels will continue to rise exponentially even if every single machine pumping greenhouse gases into the atmosphere were to be switched off right now. In the past, the slower pace of change gave the oceans time to "adapt" to the rising carbon dioxide levels, a bit like when you ask someone to shift to another seat by giving them a gentle nudge—nothing extreme, not pushing, just an arm to lean on. These days, we're doing the exact opposite. We're bulldozing the planet, ramming it and forcing it to move somewhere else.

Patrick and I take care of the last few things: pliers and screwdrivers tidied away, radio batteries plugged in to charge, items the others have left around the camp stored in a corner. The whiteness of the landscape at night, the absence of darkness, and the starless sky is almost desolate. I can hear the swoosh of water—a stream must've cut its way through the glaciers—but I have no idea if it's close by or far away. "Well, the river runs deep and the water is cold as ice," J. J. Cale once sang. The music I can hear is of an experimental Arctic genre.

Water released from the melting of glacial ice will reach—in some cases, flood—the coasts of the United States, South America, and China, but it won't happen at the same rate around the earth (as many believe, citing the popular bathtub model). Sea levels rise differently, depending on the location and the sea. Around half of the growth is triggered by the heat gained by the water: Liquids (like seawater) expand when they heat up, and the warmer they get, the more they expand. It follows, then, that as the temperature of the sea gradually rises, sea levels will also rise. And where the ocean is warmer (by the equator, for example) sea levels will rise more than where the sea is colder.

Another factor affecting where meltwater ends up is gravity—or, rather, changes in gravitational fields. The Greenland landmass exerts a gravitational pull on the ocean around it, literally pulling the water toward its icy shores as if under a magic spell. But as the ice mass melts and shrinks, its gravitational pull diminishes and the water around the island is "released" from the spell. In a nutshell, the melting ice causes changes in ocean circulation and distributes the corresponding sea level rise unevenly around the globe.

This interaction between the various forces of nature and how they offset each other is one of the things I find most fascinating about the world. Unfortunately, in this case, the interplay will have dire consequences.

Patrick is napping in a chair, book in hand. I nudge him and glance over at his tent, raising an eyebrow. I don't want him to fall asleep outside and end up a victim of the elements or cold. He doesn't wait to be reminded again and heads straight off for a well-deserved night's sleep. It takes a few more maneuvers to get out of his clothes—the same ones we go through in the morning, only in reverse—but with the added fatigue of the day on our backs.

I sink back into the chair again—stretch my legs out, rest my hands on my stomach, stare at the horizon. For a fleeting moment, a sense of nostalgia for the world I've left behind creeps up on me: the real world, filled with tiny everyday things. It's a light, languid sort of melancholy that vanishes when I take in the sheer physical beauty around me. I sigh. Sometimes I feel more alone in a crowd of people than I do out here, far away from everything and everyone, in an icy, boundless retreat.

My eyes are drawn to the glimmering ice before me, and I think how much I miss the night and everything associated with it: the particular sounds and emotions that come with it. It's like I'm a guinea pig in an experiment investigating how humans behave in an excess of light. Maybe I'll lose my mind, like a character in the film *Insomnia*, which, unsurprisingly, was set in Alaska.

It's hard to believe how much light there is in our cities when, in theory, they're so much darker. It's no laughing matter. According to a 2016 report, "light pollution" affects 99 percent of Europeans and Americans (in the US it has been rising by 10 percent every year). On a global level, estimates point to a 2 percent rise in annual pollution levels and show that 80 percent of the world lives under light-polluted skies.

I've lost track of the time. It could be midnight or two in the morning. I look at the sun, now sitting low in the sky, tepid, pale enough for me to look straight at it.

A column of ice no more than 3 feet (1 m) from the tent door, as big as an altar candle, falls from where it has stood immobile for most of its existence. The scene reminds me of some photos I saw in an article on light pollution in *National Geographic*, showing the glow in the night sky over Las Vegas and Los Angeles, which is visible up to 62 miles (100 km) away; or the financial district in Toronto, where thousands of birds are so blinded or confused by artificial light that they regularly fly into the windows of buildings; or even sea turtle hatchlings, who mistake artificial light reflecting off the ocean for the moon that they normally follow to orient themselves, and end up the victims of predators or dehydration. And the problem for human beings: Excess light can damage retinal photoreceptors or disrupt the circadian rhythms regulating sleep, during which a lot of processes vital to our health take place.

Which brings us back to the night. I remember it well, the real nighttime—so dark it hurt your eyes. Nights spent on the mountains of Irpinia, gazing at the stars. It comforts me to know that that place will live on inside me, or in my memory at least. I hear the zipper going up on Patrick's sleeping bag, then a sigh. He's asleep already while I continue to dream with my eyes wide open.

Epilogue

The sun drops behind the Statue of Liberty in New York Harbor. It's a spring afternoon, and it's still cold.

It's been a year now since the expedition to Greenland, and my memory of it is still warm and vivid, even as it merges with memories of other adventures and expeditions—fused together in a single sequence of images and feelings like neurons swirling around in an Arctic soup. Paolo, a student from Brescia visiting Columbia University, is sitting by my side as we drive along the West Side Highway, which is on the far west side of the most densely populated island on the planet—Manhattan. He reminds me a little of myself a few years ago: lively; playful but also attentive and diligent; ready to learn, and fascinated by a world he's just now getting to know, whispering in his ear the way the sirens did to Odysseus, "Stay, stay!" I didn't take much persuading. Only time will tell for Paolo.

We're going back to Greenland this year. I'm as excited as I was the first time. I can't wait for the new challenges and encounters, surprises, and difficulties—but most of all, for that feeling of euphoria that comes with discovering new things, looking out at the world while looking inside yourself: understanding a little more about ourselves. We need to prepare and train for it as usual, perhaps more than before. The intention is to continue our work on cryoconite holes, and to fly our drones to gather data that we will, when the time comes, compare with models and information received from satellites.

I look at my suitcase lying open on the floor. I can't help it. Every time I'm about to set off on an expedition, I think back to my childhood, to my answer when grown-ups would ask what I wanted to be when I grew up: "A Scientist"—in my head it was always capitalized—I'd say, without having to think about it. My mind goes back to when I arrived in Florence for my master's with nothing but the suitcase I was carrying and an offer of a bed at my good friend Domenico's. That's when I first felt the sense of solitude that only someone who has left their home behind can know. I recall when I first joined NASA, the nerves and the fear, the many times I wished I could be one of those rockets launching satellites, flying to infinity and then back "home" again. I see the faces of all the people I've cared about and still care about, whose smiles, whether mere glimmers or broad ones, have illuminated my path and helped me overcome the many challenges I've come up against along the way.

I look at my clothes lying quietly on the bed, awaiting a new voyage, and smile. My skin tingles with gratitude for being able to live my dream of becoming a scientist, and the great fortune I have at being able to share it. There's no getting away from it: The ice still fascinates me, even after all this time. Perhaps more than ever.

I turn to look out the window, at the skyline jammed full of skyscrapers: Even here, thousands of years ago, the ice made its mighty presence felt—silently, gouging canyons between Central Park, Harlem, and Hudson. I feel truly alive and extremely lucky. Ahead of me I have another trip and, once again, I get the chance to pit myself against the pale, icy land of shimmering blue and the sun that never seems to set, and the astonishing beauty that nature brings forth from its icy depths. Another chance to work alongside the greatest scientific minds, confer with them face-to-face on the most extreme battlefield the environment has ever confronted. Another chance to contemplate the erosion of the very same mystery that we ourselves were born out of.

One more time in the presence of the majestic ice.

Index

About the Authors

MARCO TEDESCO is a research professor at the Lamont-Doherty Earth Observatory at Columbia University. After receiving his Laurea degree and PhD from the University of Naples Federico II and the Italian National Research Council, he went on to join the NASA Goddard Space Flight Center as a postdoc and later, as a professor, became the founder and director of the Cryospheric Processes Laboratory. Tedesco has been featured in Science and frequently speaks as an expert on polar regions for *The New York Times*, *The Washington Post*, NPR, and others. He lives in New York.

ALBERTO FLORES D'ARCAIS was born in Rome and graduated from the University of Rome with a degree in philosophy. He's written for newspapers and magazines since the 1970s and has reported on hard-hitting issues like civil wars, drug trafficking, and the collapses of dictatorships internationally since the 1980s. In 2002, he was a John S. Knight Fellow for Journalism at Stanford University. He now divides his time between New York and Rome.